"The Paradigm

Or Tale?

Of Evolution"

A Christian-Scientific
Research

- Second Edition –

Julio A. Rodriguez
Chemical Engineer
Ex-*atheist*

Everyone MUST read this book!

"The Paradigm – *Or Tale?* – of Evolution

Julio A. Rodriguez

All rights reserved.

Copyright © 2009

ISBN:1448605822
EAN-13: 9781448605828

Original language: Spanish

Translated to English by: Sandra M. Alvarez

Classify as: Evolution, Creationism, Science, Conflicting Themes, Discipleship.

Printed in the United States of America.

About the Author:

Julio Alberto Rodriguez was born in the Dominican Republic in the year 1955; and in 1978 graduated as Chemical Engineer in the prestigious "Pontifical Catholic University Mother and Teacher"

For more than 14 years, he was an atheist; and in his own words tells us:

> "Since I did not believe in the existence of God, I believed in evolution; and I searched in the world of science the answers to life that my heart demanded."

> "For some time, I was looking to know why things happen the way they do; why for example, does the light of the sun continues to shine for so many years without getting exhausted; why does the force of gravity exist; the immensity of the universe and the perfection of the atom, etc.

> I started to imagine that perhaps some kind of **computer** would exist in the universe that controlled everything and it created gravity to avoid the collision of Astros; but then came the question: Who designed and programmed that computer? What mind was behind such perfection in the universe?"

In this book, and as a result to intensive inquiries on the subject of evolution, Chemical Engineer Julio Rodriguez presents his conclusions after 30 years of completing his studies and after innumerable life experiences.

Other book by the same author:

"The Missing Link – in Theology"

Also available in **Spanish**:

"El Eslabón Perdido – en la Teología"

In which are discussed and analyzed, among others, this concern: How can we know the treatment of God with each person, regardless of whether or not they are Christian, if they have been good or bad, or if they belong to this or any other religion?

"The Missing Link-in Theology" is a book that will guide you towards a great truth revealed in the sacred Scriptures and that would normally be dismissed and produces controversy even in the same Christian scope.

Read it. Your spiritual discernment will be illuminated.

Published in the year 2007, by Editorial Nueva Vida

53-21 37 Avenue, Woodside NY USA 11377

www.editorialnuevavida.net   www.eleslabonperdido.com

PROLOGUE

I have had the opportunity to know the development in the intellectual and ministerial field of Julio Rodriguez. In the working environment, I could also see the discipline and responsibility with which he gave to companies and institutions where he worked, during his professional years.

In addition, I have seen his continuous quest for knowledge, participating in national and international forums, with the purpose of being at the forefront of technology, in his area for that, in some way, to contribute in the progress and development of the country that saw his birth; and paying back even if it is in part, the entire investment which made its way in his moral and educational formation.

However, his existential emptiness was not satisfied with the knowledge and the good deeds he did until Christ transformed his heart, his perspective on life and his reason for everything created.

In this book, you are given the answers to the unknown that every person has in their formation as a rational being, and the enemy of the souls, using subterfuge by all means possible, has maintained humanity uninformed, preventing that the truth of God about creation, reach every human being.

I thank God for allowing that one person with the investigative capacity of Engineer Julio Rodriguez who was willing to investigate this myth of evolution, and I know that every reader will feel peace in his heart when reading this book, because it responds scientifically to us, validating the Word of God, clarifying the distortional information that our children receive throughout their different levels of education.

Rev. R.A. Benjamin Rodriguez L.
Agronomist Engineer and Brother of the Author

Dedication

I dedicate this book with a lot of love, to my lovely wife, Leonor; my children Carolina and Julio Jr.; my father, siblings, grandchildren, relatives and friends; and in a special way, to my partners in the Ministry, collaborators, and to all my brothers and sisters in the faith.

With all due respect, I extend this work to all the people who seek to know the truth of life and are not submissive in accepting what others want to instill in their minds; and they discern with wisdom, the absurd arguments that corrupt knowledge.

Acknowledgement

To all those people who showed their patience and love when, in one way or another were hurt and affected by my criticism and foolish comments, when I mocked God and all those who believed in a "Superior Being"; mainly my mother, who died without giving me her blessing because I had prohibited her to mention the word "God" on my life;

To all those who looked at me with eyes of mercy, in understanding the state of ignorance that I was in; and however maintained the hope that one day I would discern the truth that has been blindfolding to scientific knowledge;

To all those that have shared with me some moment of their lives and have enriched me with their true friendship: THANKS!

"When wisdom entereth into thine heart,

And knowledge is pleasant unto thy soul; Discretion shall preserve thee, understanding shall keep thee"

(Proverbs 2: 10-11. KJV*)*

The wise words serve:

To know wisdom and instruction; to perceive the words of understanding;

To receive the instruction of wisdom, justice, and judgment, and equity;

To give sagacity to the simple; to the young man, knowledge and discretion.

A wise man will hear, and will increase learning; and a man of understanding shall attain unto wise counsels.

(Proverbs 1: 2 -5. KJV)

Introduction

If there is truly a controversial and fascinating subject, that at one point in your lives affects all human beings, it is the subject of the origin of the universe and of life.

Each day the amount of people who think and proclaim themselves as the "spokespersons for others" increases.   It increases among both those who defend biblical creationism as well as those who are bastions of the Darwinian evolutionism.

In almost all countries of the world, large and small, has been insistently talking about this subject and many consider themselves responsible of defining the "truth" (or at least, in his view, has less lies or mistakes); and then seek the way of imposing their criteria to the society in which they serve.

In the Bible, we find the following statement: "Through faith we understand that the worlds were framed by the word of God, so that things which are seen were not made of things which do appear." (Hebrews 11:3)

Then, those who today speak of reason and logic as the only bases of knowledge, say that since the subject of faith cannot be reasoned, it should not be taught in schools; but that students should be taught nothing more than facts and science; the practical stuff, the "real".

However, at the end they find that what they claimed with such security to be logic and not faith, reason and not illusion, science and not fiction, ironically has come to require a degree of faith even greater than the level of faith sustained by all the religious of the world;
And, to avoid the embarrassment, they have changed the terms: It wasn't God who created the things, they say, but it was "accidental" everything that happened.

They say that "Nobody" made everything and the laws that exist in the cosmos and nature were rendered alone (although such laws are so strict and powerfully established by "no legislator", that no one can violate them).

Even in this 21st century, we realize that this "Nothing" was so wise and powerful, that even with all the supercomputers and incredible technology with which humanity counts on, we cannot untangle everything it has done, not even on earth;
Much less in everything else that we need to know about the other planets of the solar system; and then in our galaxy and then in the billions of galaxies that exist in the universe...

That "Nobody" that used "Chance" so that everything, out of nothing, would come to exist, is an insult to all thinking minds that do not want to recognize, because of pride, their limitations as being designed and created by the power of someone else

(Who made things His way and did not ask for our permission to come and live in this world, and will not ask for our permission to take us out of it).

Evolution becomes another dogma that one must believe "by faith". With only saying that "everything happened like so" and forcing others to believe it, does not prove anything.

I believe that the time is drawing near when all human beings will have access to the knowledge of the Absolute Truth; and that all deception, half-truths and lies, will disappear.

This modest investigation seeks to contribute in any way, to accelerate the arrival of this long awaited moment.

Table of Content

Seeking the Meaning of LIFE...

We know that we are alive, and yet we ask ourselves:

Where did life come from?

Why and for what do we live?

How did everything start and how will it end?

If man should die, will he live again?

Evil does Exist

Some time ago, somebody sent me this anecdote:

Germany
Start of the XX century

During a conference with many people from universities, a professor at the University of Berlin gave his students a challenge with the following question:
- "Did God create everything that exists?

A student courageously responded:
- "Yes, He created...

"Did God really create everything that exists" the professor asked again.
- "Yes sir, the young man answered."

The professor answered: "If God created everything that exists, then God made evil, because evil exists! And if we establish that our deeds are a reflection of ourselves, then God is evil!"

The young man shut up itself as opposed to the answer of the teacher, who happy, rejoiced to have proven, once again, that the faith was a myth.

Another student raised his hand and said:
- "Can I ask you a question professor?
- "Of course you may", answered back the professor.

The young man got up from his seat and asked:
- "Professor, does the cold exist?"

- "But, what kind of a question is that? It is obvious that it exists! Or have you never felt the cold before?"

The young man answered:
- "In reality, sir, the cold does not exist. According to the Laws of Physics, what we consider to be cold is really the absence of heat."

"All body or objects are feasible of study when it owns or it transmits energy; the heat is what causes that this body has or transmits energy."

"Absolute zero is the total absence of heat; all the bodies are inert, incapable to react, but the cold does not exist. We created that definition to describe how we feel when we are not warm."

"Does darkness exist?" continued the student.
The professor answered, "Yes it does."

The student responded:
"Darkness does not exist. Darkness, in reality, is the absence of light."

"We can study the light, but not darkness!"

"Through the prism of Nichols, the white light in its several colors can be disarranged with its different wavelengths; but not the darkness!"

... "How is it possible to identify how dark a certain space is? It is based on the amount of light present in that space."

"Darkness is a definition used by man to describe what happens when there is absence of light."

Finally, the young man asked the professor:
"Sir, does evil exist?"

The professor responded:
"Like I said in the beginning of class, we see crimes, violence all around the world. Those things come from evil."

The student responded:
"Evil does not exist, sir, or it does not exist on its own. Evil is simply the absence of good..."

"As in the other examples presented, evil is the definition that man invented to describe the absence of God."

"God did not create evil. Evil is the result of the absence of God in the heart of human beings."
"It is the same as what happens with the cold when there is no warmth, or with darkness when there is no light."

The student received a standing ovation, and the professor, nodding his head; stayed in silence...

The director of the University went to the young student and asked him:

"What is your name?"

"My name is **Albert Einstein**"

In my personal journey looking for the truth about evolution, I have found that the best way to call the **paradigm of evolution** is simply:

"The Tale of Evolution"

I can clearly perceive and declare that...

It is the one story that mocks science, logic, and reason.

-It **mocks** every mind that thinks and reasons.

-It **mocks** all logic and common sense.

-It **mocks** perfection and scientific knowledge.

-It **mocks** the human being and all its wisdom.

- It **mocks** sincerity; and

-It **mocks** faith, being faith and pretends not to be it...

...It's an utter Mockery!

In the following pages, you will see information that normally does not receive too much importance...but when we get to know them, a veil that has covered our eyes begins to fall and we can know the reality of life.

Once upon the time...

In no particular place, before time, space, energy, wisdom, and the order of things, that "something" exploded...

(Important Note: Even though I identify it as a "Tale", it is a sad reality that is affecting millions of people all over the world)

Famous Quotes:

"My people are destroyed for lack of knowledge"
Says **God** in Hosea 4:6[1]

"A little scientific information moves one away from God; but too much scientific information, brings you closer to him" **Luis Pasteur**, XIX Century

"Ignorance kills people; it is precise to kill ignorance"
José Marti, XIX Century

"There is no blinder person than the one who does not want to see"
-Famous Quote

"There are people who say and hear so many times a lie that they come to believe it is the truth"
-Famous Quote

My experience as an atheist

Like a majority of the people, I believed in God throughout my childhood; but in my adolescence, as I studied different subjects in college, and after studying Philosophy, I stopped believing in God in the year 1974.

I could see in awe and wonder how ample was the way of thinking of the great philosophers; and nevertheless, how many contradictions there were between them!

This brought a great frustration into my life. I started to see life as a big fraud; as if everyone wanted to deceive and take advantage of you. This also led me to trust nobody.

For more than 14 years, I was an ATHEIST. Not only did I not believe in God, but also I laughed at those who would talk to me about God and I would argue hard with them.

I hated when people would mention the word "God" to me and I would get very angry at the person who would do it.

I used to say that God was:

> "**A myth** created by weak and ignorant minds, that their intention, besides causing people to be good-behaved citizens, was to comfort those sorrowful people who, because of their lack of competence or preparation, faced the failure of their lives with the vain illusion that after death, they would live delightfully."

However, I would ask myself, "Who gave man the knowledge to grasp the chemical processes and improve the style of life of humanity?

How it is that the laws of gravity, inertia, and other laws of nature exist?

How is it that the planets were at the precise distance from the sun and between each other, so that life could be on Earth?

How could life initiate from inert things, when it has been proven time after time that a life comes from another life, not of something dead?

What was first, the egg or the hen?

I came to believe in God again in the year 1988; but it was not until the year 2008, **34 years later**, that I could understand **the motive** that drove me to becoming an atheist:

> God showed me that I became an atheist so easily because in school first and then in college, unconsciously, I had been *indoctrinated* with the "theory" of evolution.

We do not realize the following: We send our children to school so they may study, learn, and become prepared to deal with success the challenges of life. What we least expect is for them to become faced with an established system, which, using "impressive" and well planned terms, forces them to change the family and moral values that we as parents have taught them.

Then we say that our children "became rebellious", when the only thing they are doing is sincerely responding according to the indoctrination they received in their schools, which is then reinforced in college in the Biology class.

It is surprising the way evolution converts people into atheists.

For that reason, we see that, as well as the Christians celebrate the birth of Jesus Christ, the atheists celebrate the birth of Darwin.

I have with me a report published on February 28, 2008 by a newspaper from the state of Florida in the United States[2] , where it informs us that "more than 200 atheists would get together in "Fern Forest Nature Preserve" located in Coconut Creek, to celebrate the birth of Darwin in an international festival."

According to the report, an atheist of 40 years old said, "It is time for people who are in favor of reason and science to let themselves be heard"; and "Civil society should not be organized around personal beliefs."

On the other hand, in this article, it informs us that the president of the atheists of the county of Broward and co-sponsor of the "Day of Darwin" said:

"…many aspects of science are under attack"; and that "he is prepared to speak - until the subjects of intelligent design, abortion and the investigations of stem-cells, are no longer a subject of discussion ever again."

It is very important for people to express their opinions about the different topics that affect society… only that sometimes things become more violent, and conversations turn more aggressive.

According to the newspaper "WorldNet Daily", on February 19, 2008 when in the state of Florida, public hearing were being held so that many people could voice their opinions on the subject of evolution in schools[3], one person wrote in a forum the following:

> "You, religious fanatics, make us look like slow idiots for the rest of the world. Asia is going to surpass us in science and technology. You are making damage to the United States worse than any terrorist or communist regime. Your stupid medieval myths do not belong in our classes. Go to your mega-churches, and during your church services, spill out all your ignorance. Leave our children in peace. **EVOLUTION is REAL-The Bible is a MYTH!"**

More than thirty years ago, I would have given that person a standing ovation, because my comments were just like that.

On the other hand, I know something today, that in those days I did not:

- The Bible has a particular definition for atheists:

29

° "The FOOL hath said in his heart: There is no God." (Psalm 14:1)

- That the Lord Jesus had said:

° "If ye continue in my word, then are ye my disciples indeed; and ye shall know the truth, and the truth shall make you free." (John 8: 31-32)

As the years went by, and after revising many results of numerous scientific investigations that honestly have been made, I am now clearly convinced that:

The true **myth** of all the centuries is the concept of **evolution**, that says that:

"The universe formed itself on its own (out of nothing) stating that no superior mind can be behind such perfection; and that it has been evolving (developing itself) for millions of years, by pure chance".

However, along comes this ambiguity:

If **science verifies** (as it does, even though they do not want to recognize it) **that evolution is NOT possible,**

-Why do so many scientists doubt what they are proving in their own research?

-Why is it so hard for them to believe in the existence of the all powerful and eternal God, who created and holds the heavens and earth by the Word of His power, as the Bible tells us in Hebrews 1:2-3?

We all know that, depending on what source we are learning from, is how our understanding and **behavior** will be determined.

As a popular saying goes: "Tell me what you believe and I'll tell you who you are"; or the Bible: **"For as he thinketh in his heart, so is he."** (Proverbs 23:7)

Let us take note on the following: WE should all accept and believe one of the two answers, to the question:

Does a Creator Exist, Yes or No?

Then the following dilemma appears to us:

Whom will we believe?

Bible or Charles Darwin?

(In order to know why I only make the comparison with the Bible and not with any other book, please read the notes on the singularity of the Bible, in **Appendix 1**)

The Testimony
Of the
Dinosaurs

The figures of dinosaurs have invaded our generation.

They are in abundance on television programs for children, in articles we hand out, in movies and cartoons, in museums; and to top it all, schools use dinosaurs as "Evidence", not that they existed; but that evolution is real, and creation is a myth.

Normally, they try to confuse children and adults by asking them "to show where in the Bible it mentions dinosaurs."

With this, they want us to understand that the Bible denies their existence, but that the fossils declare the opposite; however, we see that **the Bible speaks of dinosaurs** thousands of years before man would use that name. (The word "Dinosaur" has only been used in the last 200 years)

In addition, the Bible mentions them as **contemporaries** with human beings (not millions of years before man lived on earth, as evolution declares).

The word **dinosaur** is a cultism coined by the British zoologist Richard Owen (1804-1892), using the Greek words δεινος, (deinos = terrible) and σαυρος (sauros = lizard); that is **"terrible lizards"**.

The Bible, instead of "dinosaur", mentions the following words:
- **Dragon**
- **Monster**
- **Leviathan**
- **Behemoth**

Let us see some examples...

- "So God created the **great creatures** of the sea..." (Genesis 1:21 NIV)

- "Praise the LORD from the earth, ye **dragons**..." (Psalm 148:7 KJV)

- "In that day the LORD with his sore and great and strong sword shall punish **leviathan** the **piercing serpent**, even **leviathan** that **crooked serpent**; and he shall slay the **dragon** that is in the sea. (Isaiah 27:1 KJV)"

- "It was you who split open the sea by your power; you broke the heads of the **monster** in the waters. It was you who crushed the heads of **Leviathan** and gave him as food to the creatures of the desert" (Psalm 74: 13-14 NIV)

- "Can you pull in the **Leviathan** with a fishhook or tie down his tongue with a rope? Can you fill his hide with harpoons or his head with fishing spears?" (Job 41:1, 7 NIV)

- "There is the sea, vast and spacious, teeming with creatures beyond number-living things both large and small. There the ships go to and fro, and the **leviathan**, which you formed to frolic there" (Psalm 104: 24-26 NIV)

- "Look at the **behemoth**, which I made along with you and which feeds on grass like an ox. What strength he has in his loins, what power in the muscles of his belly! **His tail sways like a cedar...**" (Job 40: 15-18 NIV)

- "Nothing on earth is his equal - a **creature** without fear" (Job 41: 33 NIV)

- "Can you pull in the leviathan with a fishhook or tie down his tongue with a rope? Can you fill his hide with harpoons or his head with fishing spears? (Job 41: 1, 7 NIV)

If you wish, you can read about other characteristics that dinosaurs had, from the other part of the book of Job, Chapter 41, in the Bible.

¿Evolution o Creation?

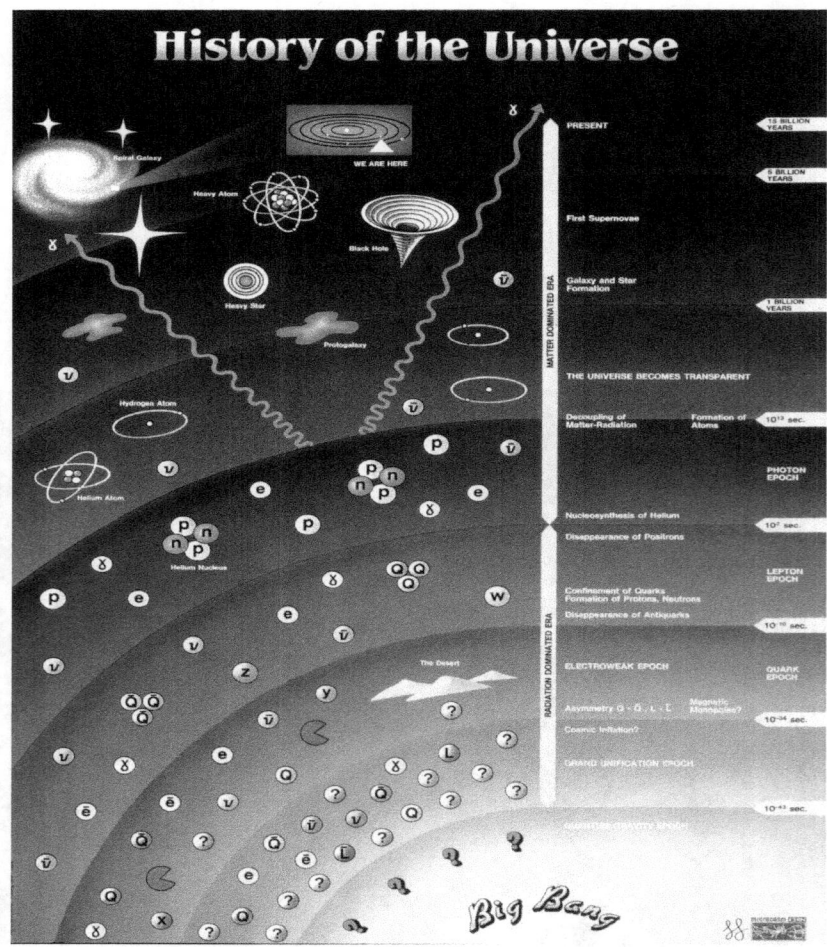

- The debate of the centuries [4] –

Jesus said:

"Render therefore unto Caesar the things which are Caesar's; and unto God the things that are God's." (Matthew 22:21 KJV)

In the Bible also it is said to us:

"Give everyone what you owe him: If you owe taxes, pay taxes; if revenue, then revenue; if respect, then respect; if honor, then honor" (Romans 13:7 NIV)

If we really reasoned with ourselves and were sincere, we all believe that all designs have a designer; and while more complex the design may be, the designer must be more intelligent!

What would the persons who have **designed**, and **constructed** the Great Hadrons Collider (LHC), think of you, if you said that it has been constructed by ITSELF, from NOTHING, and by CHANCE? [5]

The Great Hadrons Collider, located in a tunnel under earth between the border of France and Switzerland, is the most powerful and largest particle accelerator in the world.

[A hadron is a subatomic particle that experiments nuclear force. These are not fundamental particles; and are made of: Fermions called quarks and anti-quarks, and of Bosons called Gluons. The Gluons act as intermediaries for the color force that unites the quarks to each other]

It was designed to make seven trillion electron volts or (TeV) of proton beams of energy, and its main purpose is to examine the validity and the limits of the standard model of the particle physics, an actual theoretical frame of the particle physics.

More than two thousand physicists from 34 countries, from hundreds of universities and laboratories have participated in its development and construction.

This "apparatus" consists of an enormous magnet ring where millions of protons will travel at **27 kilometers** in a single direction.

Constructed by man, great works of engineering in which hundreds of scientists participate from all over the world.

The experiment, first of its type in the history of humanity, "will allow us to know new antecedents about the creation of the universe", and it will also give out different forms of energy through generations, and the solution to mortal diseases such as cancer, that would be able to be treated by eliminatory protons from cancerous cells

Another objective that scientists have from the construction of the Great Hadrons Collider is to "explore the behaviors of such resources that resemble those of the **Big Bang.**"

"One of the objectives is to find the *Higgs Boson*, the last particle that has yet to be discovered in an actual theory, called the Standard Model, and its purpose is to recreate the first one trillionth of a second emerged after the Great Explosion, that gave rise to the universe.

The apparatus cost almost **US $10,000, 000,000 (ten million dollars)** and it is designed to make particles crash with a cataclysmic force and show signs of a new physics upon impact."

I also ask you:

What do you think will think of you, if you tell that this has been made by **ITSELF**, from **NOTHING**, and by **CHANCE,** to Him who has designed, created, and maintained this…?

Let us put into perspective this dilemma:

On one side, the Bible tells us that God is the Creator of heaven and of Earth:

- *"In the beginning God created the heaven and the earth."* **(Gen 1:1)**

- *"So God created man in his own image, in the image of God created he him; male and female created he them."* **(Gen 1:27)**

The Lord says:

- *"Yea, before the day was I am he"* **(Isaiah 43:13)**

- *"I have made the earth, and created man upon it: I, even my hands, have stretched out the heavens and all their host have I commanded."* **(Isaiah 45:12)**

On the other side, in the schools, declaring they use logic and reason, teach our children:

The Universe did not begin a thousand years ago, as the religious people say; but it originated fourteen billion years ago, in a great explosion or "The Big Bang". Moreover, during the first fractions of a second, the universe was **so infinitely small and dense,** that space and time did not exist; and all of a sudden, matter appeared on the universe from said explosion.

They assure us and teach [6] in their *theory* (better to call it, *"the Tale"*) of Big Bang, that all matter, altogether with space and energy (including time), in the beginning were concentrated on one single spot, so small and dense, that to be able to study it, was necessary to formulae a "quantum theory of gravity."

Gamow baptized such singularity as "cosmic egg", and Lemaître called it: "primeval atom".

Similarly, they estimate that the **temperature** of "that" which exploded had to reach about **100 billion degrees Celsius-***more than 180 billion degrees Fahrenheit-* (In such conditions, not even an atom could exist in the way that chemistry defines them); and its **density** had to be unimaginably great (almost infinite).

After the explosion, they say, as *"that"* moved away in all <u>directions</u> (although space did not exist), energy began to transform itself slowly in matter; and in an instant: space and time were born!

Then they say that *space expanded itself and became cold,* and the formation of atoms, stars, planets, and galaxies began.

They also teach (or **"Indoctrinate"** telling) that:
"**The biological evolution,** introduced by Charles Darwin, is correct in explaining the processes of life[7]; and in reality, species transforms themselves through continuous processes, evolving through changes produced in successive generations..."

In addition, they also say "...This **degree of certainty** that **goes beyond** all reasonable doubt is what biologists refer to when they affirm that evolution is a fact..."

Also, they dare to assure that: "...The evolutionist origin of organisms is today a scientific conclusion established with a **degree of certainty** compared to other scientific concepts that are true, such as:

- the Earth's roundness"
- the molecular composition of matter"
- the movement of the planets"

To make matters worse, when Biology is studied in college, the textbook tells us "the concept of evolution is the angular stone of Biology, because it ties all the fields of sciences in one body of unified knowledge.[8]

But...

*Do we **Really** Know WHAT they are Talking About?*

Have we asked ourselves **how big** the universe is, what is it composed of, how much is the quantity and diversity of matter that exists; and how did it put itself in order?

Let us look at some scientific data:

The size of the universe is **Unimaginable**

At the speed of light (300, 000 km/sec.) or 186,000 miles per second, we would take 30,000 million years to go from one end of the universe, to the other.

Astronomers assume that the universe is composed of approximately 100,000 millions of galaxies; and each galaxy has between 100 and 1000 million stars.

The distances between galaxies are enormous.

The Andromeda galaxy is one of our neighbors and it is at 2.2 million light-years from ours; and the distance from the star closest to the Sun (the "Proxima Centauri") is of 4.3 light-years (equivalent to 40 billion kilometers).

[The light year is a unit of distance (and not of time), used in astronomy to measure great distances. It is the same as the distance traveled by the light in one year, at a speed of 300,000 km/sec.

One light year is equal to: 9,461,000,000.00 kms. (More than 9 Billion kilometers).

No material object can travel faster than light].

This way, being that the distance between the sun and the Earth is almost 93 million miles (150 million kilometers); the light of the sun travels this distance in about **eight** minutes.

Here are some other interesting facts about the sun:

- It surrounds the center of the galaxy at a speed of 155 miles (220 kilometers) per second.

- It is so big, that in its volume the Earth fits 1.200.000 times (at the same time, it is one of the smallest stars in the universe).

- In the center of the sun are consumed by nuclear fusion, 700 million tons of hydrogen every second, to produce the necessary energy to maintain life on Earth.

Our solar system is found on the arm of a spiral galaxy called THE MILKY WAY. Our solar system is located about 30,000 light-years from the center of this galaxy.

The Milky Way has about 100,000 light-years across. (Though it has 100 million stars, it is one of the smallest galaxies in the universe).

The way in which scientists have been able to study the cosmos, is through **telescopes.**

A Telescope[9] is an optical instrument or tool that allows see distant objects with much more detail than with plain sight.

It is a fundamental tool in astronomy. Every improvement in the telescope technology brings advances in our comprehension of the universe.

Thanks to the telescope, we have been able to discover many aspects of the stars and other part of the cosmos. What may seem like a white dot to the human eye in the middle of the night, seen through a telescope can be colors with details.

The Hubble Telescope[10]

The **Hubble Space Telescope** or (HST) is a robotic telescope located above the outer edges of the atmosphere at  about 370 miles over sea level, and completes a spin around Earth every 96 or 97 minutes, at about five miles per second.

In honor to Edwin Hubble, it was placed in orbit on April 24, 1990 as a collaboration project between NASA (National Aeronautics and Space Administration) and the

ESA (European Space Agency), inaugurating the Great Observatories program.

The telescope can obtain images at an optical resolution greater than 0.1 arc-seconds.

The telescopes used on Earth are being affected by meteorological factors (presence of clouds) and the luminance contamination caused by great urban establishments.

In addition, the atmosphere heavily absorbs certain wavelengths of electromagnetic radiation, especially near infrared light, diminishing the quality of the images and disabling the acquisition of spectra in certain clusters characterized by the atmospheric distortion.

The advantage in having a telescope beyond the atmosphere is, mainly, because in this way the effects of atmospheric turbulence can be eliminated; making it possible with this instrument to reach the limits of diffraction as an optical resolution.

The Hubble Space Telescope has been without a doubt, the project that has contributed the more to space discovery and technological development in the history of humanity. Great part of the scientific knowledge of Space is because of the Hubble Telescope.

Ultra-deep sight of the Cosmos

In the year 2008, the Hubble telescope turned eighteen years of working in missions and it has just finished its orbit number 100,000.

One of the wonders that it has found is called **Ultra-deep sight of the Cosmos**, in the deepest section that it can analyze; and there it located close to 10,000 galaxies.

It took 400 orbits around the Earth to make this kind of observations, from September 2003 until January 2004.

The observatory in orbit picked up a photon of light per minute from the weakest objects.

> [photon. An elementary particle that is the basic unit of light. Diccionario Manual de la Lengua Española Vox.© 2007 Larousse Editorial, S.L.]

Normally, the telescope picks up millions of photons per minute from neighboring galaxies.

It is estimated that the whole sky holds 12.7 million times more area than the ultra deep field.

To observe the entire sky would take at least one million years of uninterrupted observation

With everything that we are seeing, they still want us to believe that all the existing energy in the Universe was concentrated on a dot smaller than an atom!

You have a mind
And can reason...

...You be the judge!

Let us know
What some very
Important people
Think about this subject...

Luis Pasteur (1822-1895)[11]

French scientist that invented the process of pasteurization of milk and the vaccines against anthrax, fowl cholera, and rabies; and he became dean of the new science faculty at Lille University, where he said:

"The more I study nature, the more overwhelmed I feel by the work of the Creator. God has put in the smallest of creatures extraordinary properties, with which they can destroy the matter that has died."

Pasteur also said: "A little bit of scientific information moves one **away** from God, but too much scientific information, **draws** one close to Him."

In 1972, the Board of Education from California, asked **Werner-von Braun**, director of NASA and the father of the United States space program, to give his opinion about the origin of the universe, of life, and man.[12]

He said:

"Anyone who observes the laws and order that exist in the universe cannot help themselves to conclude that there has to be a design and a purpose behind it all."

"We feel insignificant in front of powerful forces that work in a galactic scale and before the organized design of nature that equip a small seed of ordinary aspect, with the capacity of becoming a beautiful flower."

Louis Bounoure, professor of Biology at the University of Strasbourg and Director of the Strasbourg Zoological Museum[13], said:

"Evolution is a fairy-tale for grown-ups. This theory has helped nothing to the progress of science. It is useless."

Paul Lemoine (1878-1940), Director of the National Museum of Natural History in Paris, President of the Geological Society of France, and Director of the French Encyclopedia, said[14]:

"In reality, the theories of evolution, the ones where our young students have been deceived by, constitutes of a dogma that everybody teaches; but each one, according to their specialty, the zoologist or the botanist, proves that none of the explanations they give are adequate."

"The theory of evolution is impossible. In spite of its appearance, deep down nobody believes in it. Evolution is a dogma where priests do not believe in it, but they conserve it for their devoted ones."

I have in my hands a very interesting article, titled:

"Mazur: Altenberg! The Woodstock of Evolution?"[15], written by Suzan Mazur, one of the most prestigious reporters on the subject of evolution. A number of her investigations have been published in many important areas of mass communications.

In her article, Mazur interviews 16 well known evolutionary biologists and philosophers; where they had gotten together in the month of July 2008 in Altenberg, Austria, to analyze the ways in which they can declare what would be the "Extended Evolutionary Synthesis."

These great scientists of professional status, decided to gather in what is called "The Altenberg 16", because the theory of evolution, as it is actually being taught, they say, has many failures.

They recognize that "the theory of evolution which most practicing biologists accept and which is taught in classrooms today, **is inadequate in explaining our existence.** It's pre the discovery of DNA, lacks a theory for body form and does not accomodate "other" new phenomena.

Among others, they give the following arguments:

The scientist Stanley Salthe, a natural philosopher at Binghamton University with a Ph.D. in zoology said: "Summing up we can see that the import of the Darwinian theory of evolution is just unexplainable caprice from top to bottom. What evolves is just what happened to happen."

On the other hand, developmental biologist Stuart Kauffman, who "had a breathtaking career, beginning as a medical doctor, honored as a MacArthur fellow (genius) and has worked with Nobel prize winner Murray Gell-Mann", said: "Darwin does not explain how life begins, Darwin starts with life. He doesn't get you to life."

Michael Lynch, evolutionary biologist and author of the book, "The Origins of Genome Architecture" has said, "Everyone is bantering around these terms complexity, evolvability, robustness; and arguing that we need a new theory to explain these. I don't see it."

He also said, "The big challenge is to connect evolution at the genome level, with cell development and the larger phenotypic level."

These are words expressed by evolutionary scientists, whom are sincere in recognizing that the theory of Darwin is a fallacy (though they still believe that the earth exists by casualty and not by creation).

Antonio Pardo, in his investigative work, "THE ORIGIN OF LIFE AND THE EVOLUTION OF THE SPECIES: SCIENCE AND INTERPRETATIONS"[16] **tells us:**

"...we can say that, from a scientific point of view, Darwinism is mistaken when attributing to the natural selection of the disappearance of species, because it infers particular questions in general, and this step is methodologically incorrect: the natural selection, understood as global process that regulates evolution, does not exist.

"Nevertheless, it is right when affirming that science cannot speak of degrees of perfection or of evolutionary ascent in the scale of the human being, and it would be wrong to directly translate the evolutionary succession as grades of perfection that are more elevated".

"Finally, it makes a mistake (many times deliberately or ideologically) when it affirms that evolution has no purpose nor shows progress: in evolution there is purpose but, since the extinctions are risky, it cannot be based on an inexistent "natural selection"; it is based on the causes that give origin to the new species of living beings, whom are directional; Darwinism has never tried to explain the origin of human beings, whom it gives randomly." (Pg. 568)

And he also states: "Fortunately, Darwinism is scientifically false, because she is one of its basic theses, the natural selection, because we are not attached in admitting that evolution happens by smooth changes of a population as a whole. We accept the observed reality: as we have mentioned before, it has been sufficiently proven that changes in species are clear-cut and appear in an abrupt way, we still do not know how." (Pg. 570)

Let us see a little of what Darwin says...

Charles Darwin[17](1809-1882).

Considered as the most important evolutionary scientist of the XIX century, he studied at Edinburgh University and at Cambridge in England.

In the year 1859, Darwin published the book, *"On the Origin of Species"*, in which he declared his theory of what he called, "The process of "natural selection". He taught that the variations that exist between individuals show that **"nature selects species that are better equipped to survive and reproduce."**

He also said that "The genetic variations that produce an increase in survival probabilities are **chancy** and are not caused by **God** (as the religious thought) nor by the tendencies of organisms to look for perfection (as Jean-Baptiste Lamarck proposed 1744-1829).

Some main points of the natural selection are[18]:

1. The individuals of a population vary in their form, function and behavior. Great part of these variations are hereditary; they can be transmitted from parents to children;

2. Some forms of hereditary characteristics are more adaptive to predominant conditions. They improve opportunities of survival and reproduction of the individual, and they help in obtaining food, to mate, to hide, among other aspects.

3. The natural selection is the result of the differences of the survival and reproduction among individuals of a given generation.

4. The natural selection leads to a better adaptation of the predominant conditions of the environment. The adaptive forms of the characteristics tend to become more common than other forms. In this way, the characteristics of a population change and evolve.

(See in Appendix #3, a brief biography of Charles Darwin and other relevant information)

However,
When we take his theory to be confronted with science, we find many surprises.

And this is what indeed, we will do.

For instance, did you know that, according to the laws of probability, it is IMPOSSIBLE for evolution to have ever taken place?

[Mathematically, 10ee50:1 *(10 elevated to the power of 50, to 1)* means **impossible**] [19]

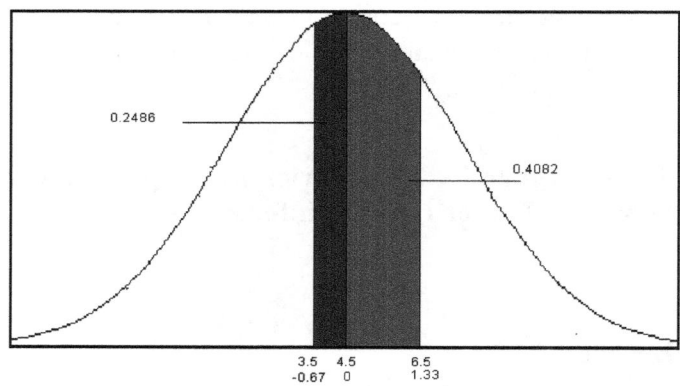

A chemist calculated the immense probabilities that existed against amino acids, so that they could combine to form the proteins necessary by a means of *"not being directed."*

Counting with time and chance, in an ideal mixture of chemicals, in an ideal atmosphere, giving it up to 100 billion years (an age 10 to 20 times greater than what evolutionists assume the Earth has), a chemist calculated [20] the immense probability that existed against the amino acids, so that they could become combined to form the necessary proteins by a means of *"not being directed"*.

The result: The probability against that the **smallest protein** being formed was more of 10 to the 67th power, to 1 (10 to the 67th: 1)

[10 to the 67[th] = 10, *000,000, 000,000, 000,000, 000,000, 000,000, 000,000, 000,000, 000,000, 000,000, 000,000]*

Generally, mathematicians conclude that, statistically, beyond 1 in 10 to 50th power (1:10ee50)[21] there is no probability that at anytime something like that will happen ("And still, this gives the benefit of the doubt!")

Two well known scientists[22] calculated the probabilities of life being formed by natural processes. They estimated that there is **less than 1 possibility in 10 to the 40, 000th** (10 to the 40,000th power is a 1 followed by 40,000 zeros) of which life could have originated by tests at random.

> *(In other words, according to the scientific law of probability, **it is impossible for evolution to have ever taken place!**)*

Not even the so-called **"The Chaos Theory"** (see documents in the reference section) has convincing arguments that indicate the possibility of evolution ever has existed.

Did you also know that evolution contradicts the laws of Thermodynamics, the most fundamental laws in physics?

THE FIRST LAW OF THERMODYNAMICS is also known as the law of the conservation of mass and energy.

According to the British Encyclopedia, 1998, this law is considered as the most important and basic of all the laws of the physics.

The first principle is a law of the conservation of energy. It states, <u>energy cannot create or destroy itself.</u> This law establishes that "the total sum of all the energy in the universe remains constant; but that a form of energy can be converted into another form of energy. The form, the size, etc. can be changed, but the total sum of the mass CANNOT be changed."

This is contrary to what the theory of evolution teaches, which says that of "zero" energy, all the energy in the universe was produced.

As Trudy and James McKee also declare[23]:
The energy of the universe is in many Inter-convertible forms: Gravitational, nuclear, radiating, calorific, mechanical, electrical and chemical. According to the modern scientific theory, the energy is the basic component of the universe"

The relation between the matter and its equivalent energy is defined by the famous equation of Einstein: $E=mc^2$

The total energy (E) in Julies (kg.m ^{2} /s ^{2}) of a particle is equal to the mass of the particle (m) in kilograms, multiplied by the speed of the light (c=300,000 kms/s) to the square potency. Nevertheless, the energy is defined habitually as the capacity to make work

THE SECOND LAW OF THERMODYNAMICS declares that "In the Universe, the quantity of energy that is available to be used...is getting exhaust. It is stating that the **Entropy** (disorder) is increasing to a maximum."

Evolution states that energy is perfecting and expanding itself; but the second law of Thermodynamics indicates exactly the opposite.

This Second Law, **contradicting what the theory of evolution teaches**, establishes that the total amount of useful energy is being reduced to a degree so that the energy is already becoming unsuitable energy and most important of everything, that this "**transformation**" is "**irreversible or irrevocable** "!

The first indication that science had about the universe **"becoming old"** and that it was becoming "wasteful" was when this second Law of Thermodynamics was formulated

In the Bible it is written:

*"Lift up your eyes to the heavens, and look upon the earth beneath: for the heavens shall vanish away like smoke, and **the earth shall wax old** like a garment..."* *(Isaiah 51:6)*

The **Entropy** is used as a measure of how far along there is disorder in a system.

In thermodynamics, the entropy is the physical magnitude that measures the part of the energy that cannot be used to produce work.

It is said that a system being highly hazardous, has high entropy; or in the same way, if the order of the universe wanted to form itself from a chaotic situation (as evolution teaches), the results would have been **greater chaos, and not order!**

As we can see, all the statements of the laws of thermodynamics contradict the theory of evolution

Still there is another very important factor, which has had a bad interpretation:
The Carbon-14.

Let us see some facts about it.

Carbon-14

- From speculation to the truth –

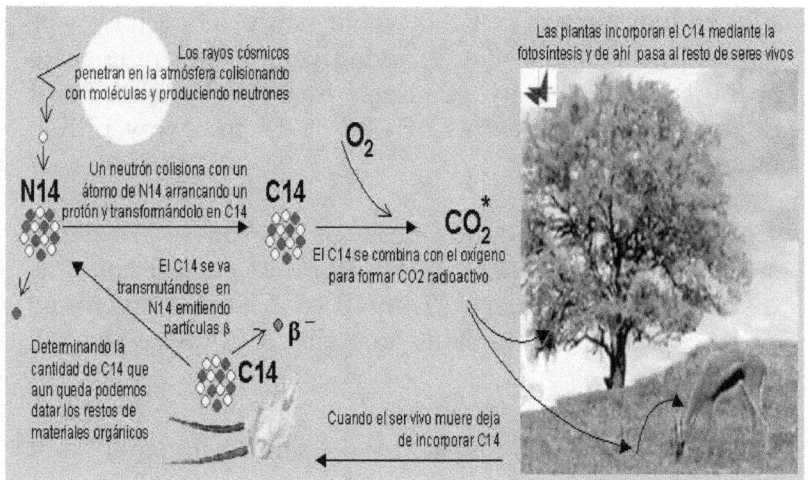

The first time that I knew something about Carbon 14, was in college; and the information I received there reinforced my atheistic beliefs and "it confirmed me" that the Bible was "a book full of fables and stories" and that "all the religious leaders were purposely deceiving the humble people"

Nevertheless, 30 years after my graduation, I have been able to know that Carbon-14 also contradicts the arguments of the evolution.

Of what this method consists?

The method of radiocarbon dating was first proposed and developed by **Willard F. Libby**, who was awarded with the good deserved **Nobel Prize in 1960.**

Bringing about numerous measurements on living matters of all kinds throughout the whole world, Dr. Libby was able to demonstrate that all living cells possess the same specific radioactivity caused by the presence of approximately 767 atoms of Carbon-14 for every billion atoms of Carbon-12.

Carbon-14 (the atomic mass of carbon-14 is about 14.003241), is a **radioactive isotope** of carbon and was discovered on February 27, 1940 by Martin Kamen and Sam Ruben. Its nucleus contains 6 protons and 8 neutrons.

Willard Libby found a value for the semi-disintegration or half-life of this isotope: 5568 years. Subsequent investigations in Cambridge produced a half-life value of 5730 years.

Because of its presence in all organic materials, carbon-14 is used in the organic specimen dating.

The method of radiocarbon dating is the most reliable one to know the age of organic remains from up to 60,000 years.

It is based on the law of exponential decay of radioactive isotopes. The isotope carbon-14 (14C) is produced in a continuous way throughout the atmosphere as consequence from the bombardment of nitrogen atoms by cosmic neutrons.

This type of created isotope is unstable; and spontaneously, it changes into nitrogen-14 (14N).

These processes of generation-degradation of 14C can be found practically balanced, so that in this way the isotope can be found homogeneously mixed with the non-radioactive atoms in the carbon dioxide of the atmosphere.

The process of photosynthesis incorporates the radioactive atom in plants, in such a way, that the quantity of 14C/12C in these, are similar to the atmospheric ones.

Animals incorporate, by ingestion, the carbon from the plants. However, after the death of a living organism, new atoms of 14C do not incorporate to weaves, and the concentration of the isotope starts decreasing while at the same time it transforms itself into 14N by radioactive decrease.

What does Carbon-14 tell us?

A very intense study has been done to 15.000 radiocarbon dating, of practically all of planet earth, and the results achieved are very different to what the students would normally think.

The information that will be presented here
is completely detailed in the study:

"TIME, LIFE, AND HISTORY IN THE LIGHT
OF RADIOCARBON DATING"
By: Robert L. Whitelaw

Also in: "Las Dataciones Radiométricas: CRITICA".
Harold S. Slusher & Robert L. Whitelaw. Edit. CLIE.
1977

These are some conclusions from the scientific studies that were made of 15,000 radiocarbon dating:

- Radiocarbon recently supports the biblical idea of Creation when it shows without a doubt a recent beginning of cosmic radiation.

- Radiocarbon supports the contemporary apparition of all forms of living matters in creation. (Modern man and animals, together with extinguished plant life and wildlife, all of a sudden they all appear ancient and without change at the same time).

- Radiocarbon supports the origin of the human race from a few predecessors in the region of the Middle East.

- Radiocarbon, on the other hand, indicates the sudden and simultaneous apparition of the animal kingdom as a whole in larger number in all parts of the world.

- The radiocarbon clearly indicates an original world in which there was profusion as much of trees as of low vegetation, and it were found present in The Poles as in the present deserted regions. (These have been greatly attested to by geology and paleontology. These also indicate that an ancient world existed singularly different in climate, location, in elevation of the continents, and perhaps an inclination in the rotation axis.)

- The radiocarbon signals a drastic change, shortly after creation, due to the cause that there was destruction in the vegetable and animal planet, but without effects in the multiplication of men.

- The radiocarbon signals with clarity a cataclysm of worldwide extension, which indiscriminately destroyed humanity, animals, and trees.

- Just as it is described in Genesis 7 and as it is confirmed in other parts of the Scriptures; and is also confirmed by preserved human traditions in all the Earth's actions, and in the evidence of worldwide geology.

- The radiocarbon supports that the date of said **cataclysm** is around **4.950 years.**

- The radiocarbon indicates a large human population, and very extended, before this cataclysm.

- The radiocarbon indicates the extended existence of extinct flowers and wildlife in the world at the time prior to the cataclysm.

(SEE in Appendix #2, the findings that gave way to these calculations)

...And as if not all of this was enough, there are at least **five** serious problems with the Theory of Evolution.

These are:

1. There is no evidence that the pre-biotic soup ever existed.

2. Transitional fossils do not exist.

3. The sudden appearance of complex forms of life.

4. It has not been proven that inert matter can transform itself into living matter by means of a natural process.

5. There are no mechanisms that can be accepted as valid.

Considering of great interest, I am going to transcribe a part from the study:

"SCIENTIFIC PROBLEMS WITH THE THEORY OF EVOLUTION OF THE SPECIES" [24]

<<
1. There is no evidence that the pre-biotic soup ever existed

There is an increasing body of evidence that indicate that the **primitive** terrestrial atmosphere **had oxygen**; therefore could not be composed of the same materials that Oparin, Haldane, and others propose.

The oxygen would destroy these prebiotic chemicals when reacting with them. Dr. Robert Shapiro, an evolutionist and biochemist, has a whole chapter titled, "The Spark and the Soup" in one of his books, which he deals with the subject of "The Myth of the Prebiotic Soup." [25]

Drs. Thaxton, Bradley, and Olsen have summarized this problem in the following manner:

"...in the atmosphere and in the various water basins of the primitive earth, many destructive interactions would have so vastly diminished, if not altogether consumed, essential precursor chemicals, that chemical evolution rates would have been NEGLIGIBLE.

The soup would have been TOO DILUTE for direct polymerization to occur. Even local ponds for concentrating soup ingredients would have met with the same problem.

Furthermore, NO GEOLOGICAL EVIDENCE INDICATES AN ORGANIC SOUP, even a small organic pond, ever existed on this planet.

It is becoming clear that however life began on earth, the usually conceived notion that life emerged from an oceanic soup of organic chemicals is a MOST IMPLAUSIBLE HYPOTHESIS. We may therefore, with fairness, call this scenario "THE MYTH OF THE PREBIOTIC SOUP." [26]

2. Transitional fossils do not exist

Scientists agree that we have fossils of the major types of plants and animals that are most important for study.

However, since the theory of evolution states that change from one type of plant or animal to another type, occurs very slowly, it is completely logical to think that many fossils of transitional or intermediate life forms exist.

For example, according to the theory of evolution, reptiles became birds over a long period of time. We should, therefore, find fossils of several animals between reptiles and birds. But, what is it that we have actually excavated?

We have found a lot, but **nothing** that can be determined as transitional fossils.

Darwin was aware of this problem when he said:
"Geology assuredly does not reveal any such finely graduated organic chain; and this, perhaps, is the most obvious and gravest objection which can be urged against my theory." [27]

However, Darwin thought that future excavations would find these transitional fossils. However, what has been found in more than 120 years?

It is best to let the scientists speak.

For example, evolutionist and paleontologist David Raup, Ph.D said:

> "Darwin...was embarrassed by the fossil record of his time...we are now about 120 years after Darwin and the knowledge of the fossil record has been greatly expanded."

> "We now have a quarter of a million fossil species, but the situation hasn't changed much...WE HAVE EVEN FEWER EXAMPLES OF EVOLUTIONARY TRANSITION THAN WE HAD IN DARWIN'S TIME" [28]

> "(And various ones that were thought to be transitional, were later on discarded)."

3. The sudden appearance of complex forms of life

We could summarize the evolutionary position on geology and life forms as follows:

1. The earth's surface is composed of several layers, with the oldest at the bottom moving up to the youngest layer on top.

2. Since the simplest life forms are the oldest they appear in the lowest layer (more deep) and gradually change through the layers to the complex life forms in the top level(s).

This means that the ancestral form of each life form should be in the layer below it and the lowest layers should have the "simpler" life forms. **However, this is not what the geological studies tell us.**

Let the scientists illustrate this to us:

Fred Hoyle, Ph.D and Chandra Wickramasinghe, Ph.D said:

"The problem for biology is to reach a simple beginning... the tendency is to imagine that there must have been a time when simple cells existed but when complex cells did not... **this belief has turned out to be wrong...**

Going back in time to the age of the oldest rocks... fossil residues of ancient life-forms in the rocks do NOT reveal a simple beginning."

"Although we may care to think of fossil bacteria and fossil algae and micro fungi as being simple, compared to a dog or horse, the information standard remains enormously high.

Most of the biochemical complexity of life was present already at the time the oldest surface rocks of the Earth were formed." [29] (Note the use of the word "belief")."

4. It has not been proven that inert matter can transform itself into living matter by means of a natural process

The theory of evolution states that simple things became complex; that non-living chemicals (macromolecules) became living cells by chance, and from that point, little by little, they began evolving into living cells with DNA.

Is that even possible? Do scientists observe anything like that today that proves this type of "miraculous transformation"?

Can we even **synthesize living matter** in a laboratory using high-tech lab equipment, computers, etc., and lots of design?

The answer is no, no, and no.

Fred Hoyle, Ph.D. and Chandra Wickramasinghe, Ph.D., both evolutionists, explain why this is not possible.

"Life could not have had a random origin... the problem is that there are about 2000 enzymes, and the probability of obtaining them all in a random trial is only one part in 10 elevated to the power of -40 thousands (10ee40000:1) an outrageously small probability that could not be faced **even if the whole universe** consisted of organic soup."

"If one is not prejudiced either by social beliefs or by a scientific training into the conviction that life originated on the Earth, this simple calculation wipes the idea entirely out of court..."

"The enormous content of even the simplest living systems... cannot in our view, be generated by what are often called "natural" processes...

For life to have originated on the Earth it would be necessary that quite explicit instruction should have been provided for its assembly."

"...There is no way in which we can expect to avoid the need for information, no way in which we can simply get by with a bigger and better organic soup, as we ourselves hoped might be possible a year or two ago." [30]

Hubert Yockey, Ph.D, an expert in molecular biology, information science and mathematical probability (and an evolutionist), stated:

> "the building blocks...do not spontaneously make proteins, at least not by chance. The origin of life by chance in a primeval soup is impossible in probability...A practical person must conclude that life didn't happen by chance."[31]

Nobel Prize winner, strong evolutionist and biochemist, Francis Crick, Ph.D, recently concluded:

"An honest man, armed with all the knowledge available to us now could only state that in some sense, the origin of life appears at the moment to be **almost a miracle**, so many are the conditions which would have had to have been satisfied to get it going." [32]

To summarize this point, I would like to quote Dr. Michael Denton again from a chapter he titled "The Puzzle of Perfection":

> "The intuitive feeling that pure chance could never have achieved the degree of complexity and ingenuity so ubiquitous in nature has been a continuing source of skepticism ever since the publication of the "Origin of Species";

And throughout the past century there has always existed a significant minority of first-rate biologists who have never been able to bring themselves to accept the validity of Darwinian claims..."

Perhaps in no other area of modern biology is the challenge posed by the extreme complexity and ingenuity of biological adaptations more apparent than in the fascinating new molecular world of the cell.

To grasp the reality of life as it has been revealed by molecular biology, we must magnify a cell a thousand million times until it is twenty kilometers in diameter and resembles a giant airship large enough to cover a great city like London or New York.

What we would then see would be an object of unparalleled complexity and adaptive design.

On the surface of the cell, we would see millions of openings like the port holes of a vast space ship, opening and closing to allow a continual stream of materials to flow in and out.

If we were to enter one of these openings, we would find ourselves in a world of supreme technology and bewildering complexity...

Is it really credible that random processes could have constructed a reality the smallest element of which -a functional protein or gene- is complex beyond or own creative capacities, a reality, which is the very antithesis of chance, which excels in every sense anything produced by the intelligence of man?" [33]

5. There are no mechanisms that can be accepted as valid.

An expert in Radiation and mutation, Dr. H.J. Muller had said:

> "There is no single instance where it can be maintained that any of the mutants studied has a higher viability than the mother species...
>
> A review of known facts about their ability to survive has led to no other conclusion than that they are always constitutionally weaker than their parent form or species; and in a population with free competition they are eliminated...
>
> Therefore they are never found in nature (e.g., not a single one of the several hundreds of Drosophila mutations), and therefore, they are able to appear only in the favorable environment of the experimental field or laboratory..." [34]

Can natural selection or mutations explain the millions of genetic changes that would have to take place exactly at the same time for a reptile to become a bird?

Does this hypothesized change in lungs (and feathers, etc.) even agree with what we know is true about the nature of mutations?

Evolutionist Dr. Pierre-Paul Grasse, former president of the "French Acadamie des Sciences" and the scientist who held the Chair of Evolution at the Sorbonne in Paris for twenty years has clearly stated the problem:

> "The opportune appearance of mutation permitting animals and plants to meet their needs seems hard to believe.
>
> Yet the Darwinian Theory is even more demanding: a single plant, a single animal would require thousands and thousands of lucky, appropriate events.
>
> Thus, **MIRACLES** would become the rule: events with an infinitesimal probability could not fail to occur. ...
>
> Certainly, there is no law against dream wide-awake, but science must not indulge in that luxury." [35]

Trudy and James McKee said it best in their book, "Biochemistry: The Molecular Basis of Life"[36]:

> "Life has demonstrated to be much more complex than what the human imagination could have conceived. The structure of cells is the case in question.

> Cells are not bags of protoplasm that scientists imagined around a century ago; but they are complex and dynamic structures...

> The scientists who work to understand the physical reality of the natural world, are frequently surprised at the sophistication that are even the simplest of organisms."

Are you asking yourself the following:

Why has then evolution converted itself into something so universally acceptable; why does that **"spirit of absurd denial"** have so much power and why is there so much hostility against the Bible?

Do you believe that everything came from nothing, without the intervention of God the Creator?

So much perfection and complication,
Does it have something to do
with **your** soul?

<u>Let us look at a recent news event:</u>

In the schools of Florida, evolution will be taught as "theory"[37]

M. CAPUTO/ Miami Herald- Thursday, August 14, 2008. TALLAHASSEE

> "For the first time, the evolution of the species will be taught in a clear and explicit way in the **schools of Florida**, after the Board of Education of the state approved on Tuesday a series of new science rules that mentions the word "evolution."
>
> With the new standards, it will be required that teachers teach evolution and natural selection starting from the sixth grade; and beginning in the **ninth**, they will teach "the evolution of hominids from their first ancestors", "Genetic deviation", and "movement of genes."

What does this **"progress"** mean?

Let us remember that less than a century ago, we did not think in this manner...

In the year 1925 there was a famous judicial case in the United States, known worldwide as the "Scopes Monkey Trial", for which teacher **John Scopes** was judged, and accused of teaching the theory of evolution to his students in **Dayton, Tennessee**.

It was a famous ruling because he became faced with agnostic believer Clarence Darrow, defender of Scopes, the most famous trial attorney in North American history, and lifelong Presbyterian Williams Jennings Bryan.

Consequently, in 1925, the House of Representatives of Tennessee approved, unanimously, a law that proclaims:

> "In the Universities or normal schools or any state-school that is being financed entirely or partially with funds from the State of Tennessee, it is unlawful to teach any theory that denies the story of the divine creation of man as taught in the Bible, and to propagate in its place that man has descended from a lower order of animals." (1925)

The Bible says in Romans 1: 19-22:

> "Because that which may be known of God is manifest in them; for God hath showed it unto them. For the invisible things of him from the creation of the world are clearly seen, being understood by the things that are made, even his eternal power and Godhead; so that they are without excuse."

> "Because that, when they knew God, they glorified him not as God, neither were thankful; but became vain in their imaginations, and their foolish heart was darkened. **Professing themselves to be wise, they became fools...**"

We have been talking about the
Immensity of the Universe

And what of the small things?

Is there anything interesting to analyze?

Let us see...

The atom and matter:

When we enter into what matter really is, we will find ourselves with its foundation: **The Atom**, which is the steadiest, smallest particle that composes matter.

We can prove, with astonishment, that, as it happens with the Milky Way Galaxy, where the sun with all the planets rotating around, and next to the millions of stars that shape it, though all the turning being done, they do not clash or leave their orbits; this also happens at the atomic structural level.

The electrons, in the same manner as the planets, rotate around the nucleus of the atom, and they never crash with one another.

The atom is formed by a nucleus and a cover.

In the **nucleus,** we find two types of particles: The protons, which are positively charged particles; and the neutrons, which do not possess any electrical charge, they only possess mass.

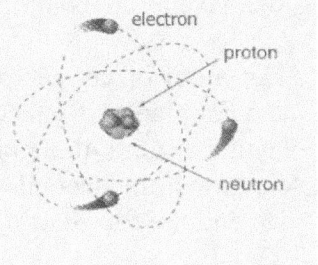

On the other hand, in the **cover** we find the electrons, which are only negatively charged particles that are found in constant movement around the nucleus.

The major part of the mass of an atom is concentrated in the nucleus, formed by protons and neutrons, both known as **nucleons**, which are **1836 and 1838 times** heavier than the respective electron.

Modern use of electrons. The electrical current

We understand as electric current to the flow of electrons that circulate through an electrical conductor, whereas the degree of conductivity of an element is given by the amount of electrons of the last orbit of the atom.

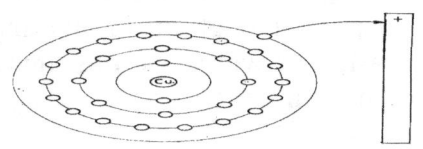

There are conductive and insulating materials from the electrical current, depending on the atomic structure of its last orbit or outer layer.

We have as an example, copper, which is a conductor.

The atom of copper possesses 29 protons in the nucleus and 29 planetary electrons that rotate in orbit inside four layers around the nucleus. The first layer contains 2 electrons, the second layer contains 8, the third layer has 18, and the fourth or most external one, has 1 electron.

Here we have represented the way in which the atomic electrons of copper flow, to create the electrical current:

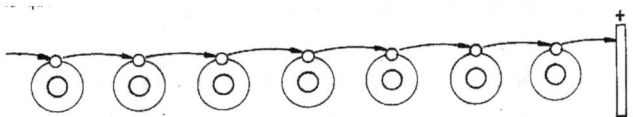

However, the modern use of electrons is not only confined to generate the electrical current; but another area of science has developed, called **Electronics**, which is responsible for the great technological changes that we are seeing in our days (satellites, computers, cellular and all of a range of digital products).

We have seen why the atom is the smallest and steadiest particle that composes matter. Now let us talk about **Matter**.

Matter is everything that takes up space in the Universe.

Matter is everything that is formed from atoms or molecules and with the property to be found in states of solid, liquid, and gas.

Let us take examples such as stones, wood, bones, plastic, the air and water. Everything we can see and touch is constituted from matter.

Matter has Mass and Volume (in other words, it can be weighed on a balance and occupies a place in space).

The objects or bodies that we see and touch can be made up of one or various types of matter (or materials).

At a structural level, all the particles that make up matter are governed by **forces** that keep them acting the way they do.

 The simplest nucleus is **hydrogen**, formed only by a proton.

The nucleus of the following element in the periodic table is **helium**, which is formed by two protons and two neutrons.

The quantity of protons contained in the nucleus of the atom is known as the atomic number.

Oh, the wisdom of the person who invented the atoms. That person gave them the capacity to combine themselves with other atoms to form diverse manifestations of matter.

Those who have study at school normally are used to see the diagram that summarizes all the types of matter, which are known until nowadays. It is what we call:
The Periodic Table of the Elements

Unions of Atoms:

Let us talk about **unions of atoms**, to form what we call **Molecules**.

Let us take as an example, **Water**.

Water is a wonderful and unique substance. In the ancient times, it was considered that water was an element; but later on, it was discovered that it was a compound formed by two atoms of hydrogen and one of oxygen.

It is the universal solvent, transparent, liquid, and has no odor, color, nor taste.

Everyone uses water as a natural resource.

Some of its uses are to drink, cook, clean; for recreational activities such as swimming, row a boat, fish, etc.

It is an important element of transportation and among other things is used to produce energy.

About three-fourths parts of the surface of the Land is covered with water.

Water is one of the most used and important resources of the planet. In its liquid state, we usually obtain it from the rain, fountains, streams, rivers, and lakes. As vapor, water is also found in the air, where is usually condensed for the formation of clouds.

Due to the water cycle, the water supply of our planet is constantly in motion, from one place to another and in one way to another. Everything on earth would suffer deterioration if the water cycle did not exist!

Earth is a place with a lot of water.

About 70 percent of the planet's surface is covered with water.

The importance of water in life can be understood if we refer to the **functions carried out by organisms to stay alive.**

In the functions that allow organisms to handle energy to synthesize and break down compounds, water plays an indispensable role in all this.

Likewise, the organic compounds, source of energy, are transported through the water.

The waste products of the organisms also use water as a vehicle. We could say that **any metabolic activity is intimately linked to the molecule of water.**

Moreover, the organisms establish intimate and transcendent relations with the environment.

Water, thanks to its heat capacity, plays a very important role in thermal regulation of the climate, making the changes less rough, than they would be if no water existed.

Inside the organism, water, also has this important function: to regulate the temperature.

The release of water vapor as sweat or panting is vital for the conservation of the body temperature.

88

Almost 75 percent of a person's weight is water.
How much water is on (and within) the Earth?

Here are some figures:

- The total water supply of the world is equivalent to **326 million cubic miles** (a cubic mile is an imaginary cube -a square box- that measures one mile on each side).

- Close to 3,100 cubic miles of water, mostly in form of water vapor, is found in the atmosphere at any one time.

It is found to be distributed in the following order:
(Data in Cubic Kilometers)

Oceans	1, 321, 000, 000	97.24%
Ice Caps	29, 200, 000	2.14%
Groundwater	8, 340, 000	0.61%
Freshwater	125, 000	0.009%
Inland Seas	104, 000	0.008%
Humidity of Earth	66, 700	0.005%
Atmosphere	12, 900	0.001%
Rivers	1, 250	0.0001%
Total: 1, 360, 000,000		100%

Source: Nace, Encuesta Geológica de los Estados Unidos, 1967 y El Ciclo Hidrológico (Panfleto), U.S. Geological Survey, 1984]

In other words, the total volume of water of the planet is equivalent to **332 million cubic miles**.

(One cubic mile of water is equivalent to
more than one trillion gallons)

(By the way, that is a lot of water, don't you think?)

One curious question:

Do you know how many

Molecules and atoms

Are in a **single drop** of water?

Here are some useful scientific data:

A MOLE (symbol MOL): is the quantity of a substance that contains 6.02×10^{23} units, or in other words, the **Avogadro's number** for particles.

1 mole of water = 6.022×10^{23} molecules of water.

MOLAR MASS: one mole of a substance is equal to the **substance's atomic weight or molecular weight**, in *grams*.

1 Molecule of water is formed by:
    2 moles of H + 1 mol of O = 1 mole of water

2 grams of H + 16 grams of O = 18 grams

As the molar mass of a mole of water is 18 grams, we then have:
    1 mole of water = 18 grams
              = 18 grams of water contains
              6.022×10^{23} molecules of water

1 drop weighs 0.2 grams

1 molecule of water contains 3 atoms (2 de hydrogen y one of oxygen)

Therefore:

The answer is:

    $\equiv 0.2 \times 6.022 \times 10^{23} / 18 = \mathbf{6.9 \times 10^{21}}$ Molecules.

In other words:

In one drop of water, there are:

6,900, 000,000, 000,000, 000,000 molecules

(Six thousand nine hundred trillions of molecules);

Equivalents to 20,000, 000,000 000,000, 000,000 atoms

(Twenty Thousands Trillions of atoms)

Can you imagine that, if to form a drop of water must get together **twenty thousands trillions of atoms,** the evolutionists wants make us think that of something **smaller than an atom, all** the universe was formed?

We see that this idea is not logical, neither reasonable, nor scientist… It is PURE FAITH.

How many atoms and molecules are then found in…?

- o   A cup of water?
- o   A pool?
- o   A lake?
- o   The whole earth?
- o   Etc…

The **absurd denial** to such perfection and complication, does it have something to do with **your soul?**

As the Bible tells us:

> "The instruments also of the churl are evil: he devises wicked devices to **destroy the poor with lying words.**" (Isaiah 32:7)

Nuclear Fission and Fusion:

Apart from having an atomic combination that produces a vital element to life, we have infinite atomic combinations that produce all types of matter.

In addition, we can learn and utilize the atomic forces that act in diverse elements.

One of the most important ones is **Uranium**.

(Core of Uranium)

Let us talk a bit about Nuclear **Fission**

Nuclear fission is the division of the nuclei of a heavy atom into lighter nuclei, which generally is accompanied by great releases of energy.

There exists two ways in which we can use nuclear energy to convert it into heat: **nuclear fission**, in which an atomic nuclei is subdivided into two or more parts;

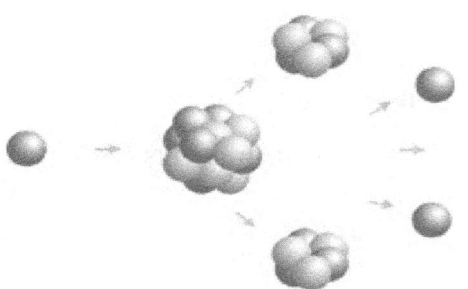

And there is **nuclear fusion**, which is the combination of at least two nuclei to form a heavier nucleus.

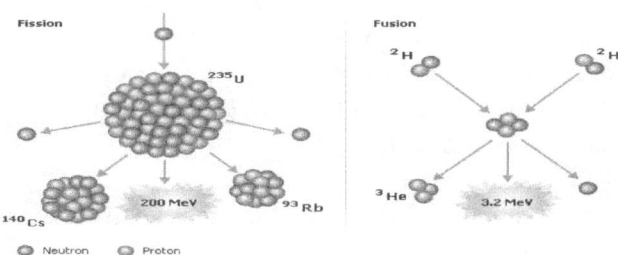

Nuclear Fission, as other things in this life, has constructive and destructive uses.

Destructive use of Nuclear Fission

Nuclear fission, in spite of being highly productive (energetically speaking), is also very difficult to control, as we can see in the **Chernobyl** disaster, and in the bombings of **Nagasaki and Hiroshima.**

Let us look at the good uses of nuclear fission:

Great part of the existing nuclear power stations at present are based on fission reactors, using a **composed combustible uranium** among a 3.5% and a 4.5% of U-235 and the rest of U-238 (This isotope is known as **enriched uranium**).

The nuclear chain reaction that generates energy is produced when a Uranium-235 nucleus is divided in two or more nuclei by the collision of a neutron. In this way, the released neutrons collide again forming a chain reaction.

In the nuclear power plants by fission, the heat of the reactions generates water vapor, which, when passing through a system of turbines, generates electricity that can be transmitted to the electricity network.

Three Mile Island Nuclear Power Station. Image By US Government

Being honest, we should ask some questions to all those who consider themselves evolutionists:

- Where does the space from the universe come from?

- Where does matter come from?

- Where do the laws of the universe (gravity, inertia, etc.) come from?

- How could matter form itself into such perfection?

- Where did the energy that organized all matter come from?

- The first cell, able to reproduce sexually, with whom did it reproduce?

- How, when, and where did human beings evolve his/her five senses, emotions and feelings?

- How did humans begin to reason, think, discern, and speak?

What Evolved First?

The DNA or RNA that carries the messages of DNA to the different parts of a cell?

The digestive system, the food to be digested, appetite, the ability to find and eat the food, digestive juices, or the resistance of the stomach to these juices?

The lungs, the mucus that protects them, the throat, or the perfect mixture of gases that our lungs breathe?

What evolved first, How, When, and Why: the flower or the bees that pollinates it?

We know that vertebrate animals have an internal skeleton formed by the articulation of bones, which are linked to one another. The skeleton maintains and gives form to the body; protects certain sensitive organs and serves to sustain the muscles.

We ask, then:

What evolved first, How, When, and Why?
The bones, ligaments, tendons, or the organs that would be protected?

Etcetera, etcetera, etcetera…

If somebody asked me how I could make **a silhouette** have **life,** develop, multiply itself, have knowledge of the things that surround it, and protect itself, etc.?

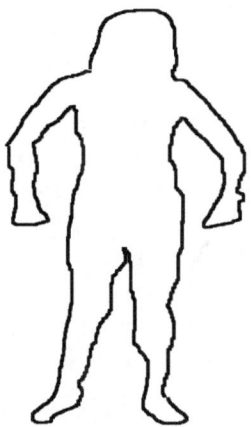

Let us see what I would say:

Because I am sure that an answer such as: "Let it form itself by chance" wouldn't be a very convincing answer, Let us see what I would say:

I would prepare a system that would sustain it (the skeletal system)

I would have to make more than two-hundred bones, about one-hundred joints, and more than six-hundred and fifty muscles, for them to **act in a coordinated manner**.

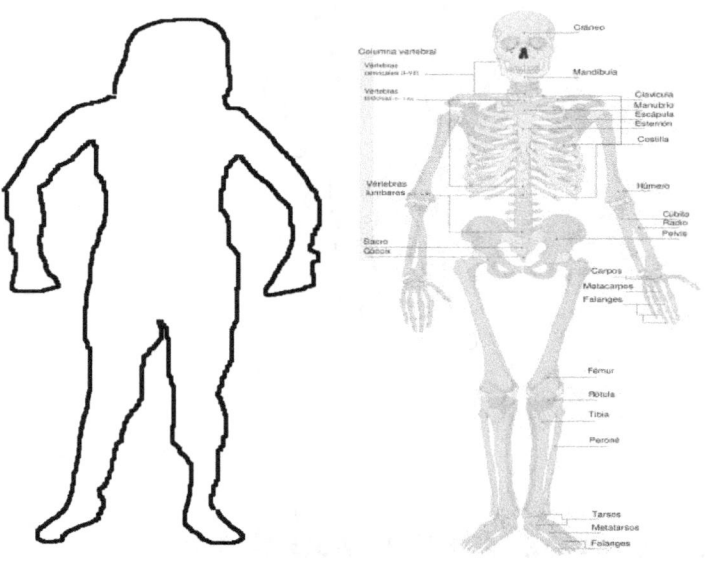

It is because of the collaboration between those bones and muscles that it is possible to maintain posture and perform multiple actions.

I would try to make the bones become united without welding, and may be fitted to accomplish the necessary functions.

I would prepare a system that allows her to move and make forces and other actions. (The muscular system)

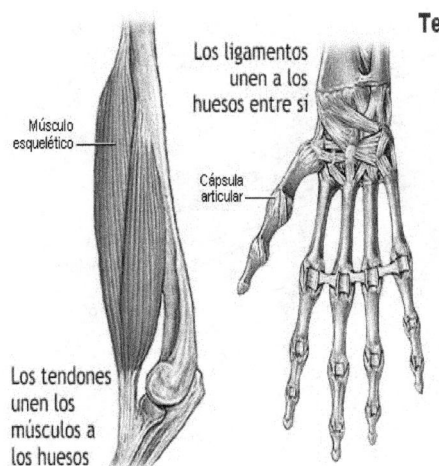

Tendones y ligamentos

Los ligamentos unen a los huesos entre si

Músculo esquelético

Cápsula articular

Los tendones unen los músculos a los huesos

Los tendones son tejido conectivo fibroso que une los músculos a los huesos. Pueden unir también los músculos a estructuras como el globo ocular. Los tendones sirven para mover el hueso o la estructura, mientras que los ligamentos son el tejido conectivo fibroso que une los huesos entre sí y generalmente su función es la de unir estructuras y mantenerlas estables.

In this manner, I would make the system (called skeleton) to sustain the organism and protect the delicate organs, such as the brain, the heart, or the lungs.

At the same time, it could serve as a point of insertion to the tendons of the muscles; and would make the bones to unite among themselves through ligaments.

Muscles have to be the engines of movement. A muscle would be a beam of fibers, whose most prominent ownership would be contractility.

Thanks to that power, the package of muscle fibers would contract when it receives an adequate order.

When it contracts, it would shorten and dump itself from the bone or from the subjugated structure. When the job would finish, it would recover its position of rest.

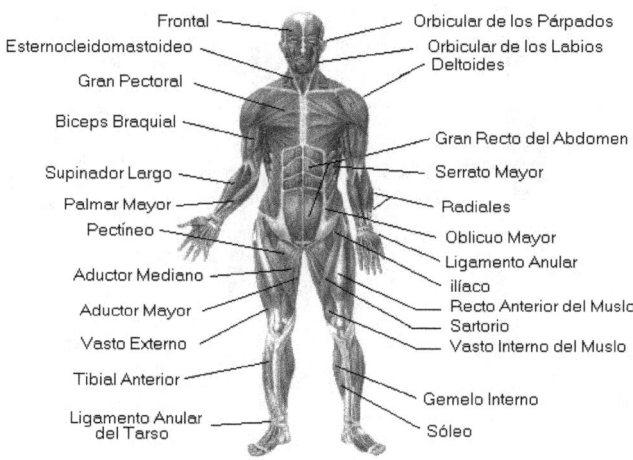

I will now put colors. I will make the fluted muscles red, they will have a fast and voluntary contraction and they will insert themselves in the bones through a tendon. For example, the ones from chewing, a trapeze, that will sustain the head upright, or the twins in the legs that will allow them to be on their heels.

For its part, I would make the smooth muscles the color white; those who have damaged tubes and pipes will have a slow and involuntary contraction.

They will eventually meet, for example, by covering the digestive tract or the blood vessels (arteries and veins).

The heart muscle should be a special case, because it is a fluted muscle, of involuntary contraction.

It would have to cover about 650 muscles of voluntary action. Such muscular wealth would allow thousands of movements. [38]

I would also make a system (the nervous system) that would help it to move or shift around[39].

Organs that transmit and process all the information that enters it, from the five senses, allowing it to move, adapting itself to the external environment, and carrying out intellectual activities, would form this system.

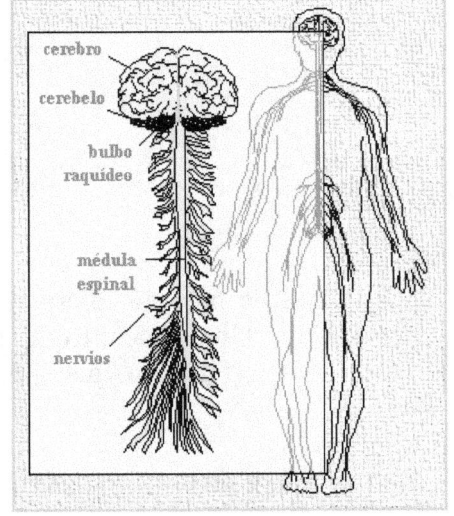

Such a system would allow it to capture the characteristics of the environment, and make an adequate mental representation of the external reality and predict the impact and the actions of the external developments, but its function would not be limited only to that, it would also receive stimulus from all the internal organs.

That is why I would have to make it a peripheral nervous system, which would be in charge of roaming the body through the nerves, receiving and transmitting the stimuli to the central nervous system.

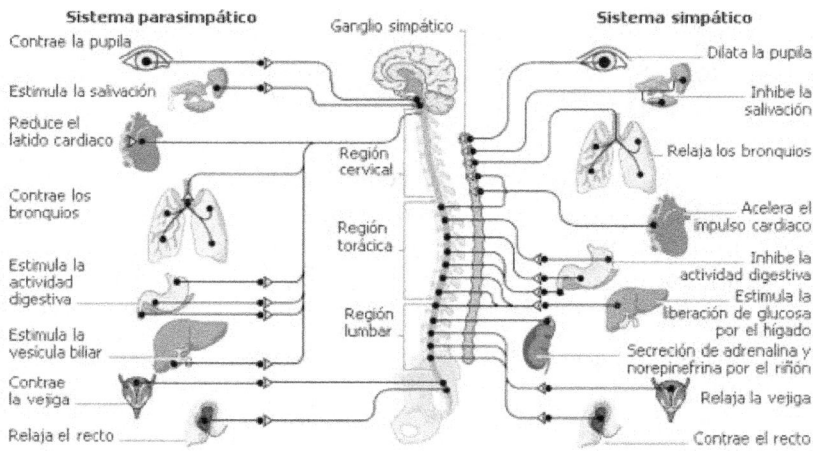

This would be busy in interpreting the stimuli and act accordingly. Give orders to the muscles and the glands to fulfill their functions according to the needs of the body.

Oh, one more fact.

Since the cells of the nervous system would be very sensitive as they could not reproduce, I would have to protect them in a special way: with the skull and spine.

And How will I make it develop and operate properly?

I will make a system that allows it to take the necessary nutrients of the things that are outside, towards the inside cells of the whole body.

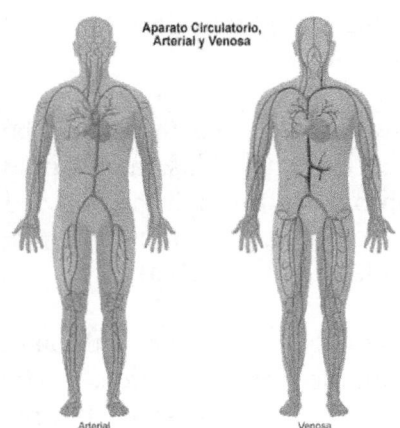

Aparato Circulatorio,
Arterial y Venosa

Arterial   Venosa

For this I will need to make a Circulatory System (vascular and lymphatic) [40]. The vascular system, **also called circulatory system**, would have the vessels that transport blood and lymph through the body.

The arteries and the veins would transport blood throughout the body, with the aim of supplying oxygen and nutrients to the body tissue and eliminating the waste from the tissue.

The lymphatic vessels would carry lymph fluid (a clear colorless liquid that contains water and white blood cells).

The lymphatic system would help to protect and maintain the medium liquid of the body, through the filtration and drainage of the lymph from every part of the body.

The vessels of the circulatory blood that I would have to do would be:

The arteries: blood vessels that would transport the oxygenated blood from the heart to the rest of the body.

The veins: blood vessels that would transport the blood of the body to return to the heart.

The capillary vessels: tiny blood vessels that would be found between the arteries and the veins, and would distribute the rich blood in oxygen through the body. I would create a bomb that would be able to maintain the blood in circulation. It would be called: **Heart.**

The blood that would come out from the heart through the arteries would be saturated with oxygen. The arteries would then divide into branches, each time becoming smaller to take the oxygen and other nutrients to the cells of the tissues and the organs of the body.

As the blood travels about aimlessly through the capillaries, the oxygen and other nutrients would be introduced into the cells, and the waste from the cells would displace to the capillaries.

As blood flows out from the capillaries, it would be transported through the veins, which are every time increasingly large, so they could return it to the heart.

In addition to maintaining the blood and lymph in circulation throughout the body, the vascular system would act as an important component to other corporal devices, for example:

And so that it may breathes, I will do a respiratory System.

As the blood flows through the capillaries of the lungs, there will be an exchange of carbon dioxide to oxygen. Carbon dioxide would be expelled from the body through the lungs, and the blood would distribute oxygen to the body tissues.

Hummmm.

I think I need to elaborate better on how the respiratory system would have to function.

The respiratory system is the one in responsible for carrying out the exchange of gases between the air and blood.

I would have to prepare the following:

1. Airways: will lead the outside air to the lungs and vice versa.

1.1 Nostrils: are the two holes that I will make in the nose. In them, the air will be filtered, heated, and moistened.

1.2 Pharynx: will form part of both the respiratory and digestive tract: it will communicate with the larynx and esophagus. It will have the same mission as the nostrils.

1.3 Larynx: in its interior, I will put the vocal chords, whose vibration with the passage of the air will produce the voice. When it swallows food, the larynx will stay closed by a kind of tab called the epiglottis.

1.4 Trachea: would be a long tube that possesses incomplete gristly rings in form of a C that always keep it open. It will be placed in front of the esophagus.

1.5 Bronchi: will be the two tubes in which the trachea divides. It will penetrate the interior of the lungs where it will ramify repeatedly, forming the bronchioles. The inside wall will possess cilium (a kind of hairlike projection that vibrates) and mucus to filter the air and trap the particles that it leads in suspension.

2. Lungs: Will be two sponge like masses with a fabric of double wall called **pleura**, with a thin layer of liquid between the two, which I would have to make to soften the respiratory movements.

The right lung will be divided into three lobes and the left in two. They will consist of the bronchioles that will be repeatedly divided into different branches, each time becoming thinner and will be completed in some air sacs called **alveoli**, coated with blood capillaries.

Lung Ventilation This is what I would call the entrance and exit of air in the lungs. It will consist of two respiratory movements; inspiration and expiration.

1. Inspiration: Will be produced by the contraction of the diaphragm (descending) and of the muscles that elevate the ribs. This would provoke an increase in the chest cavity that allows the entry of air in the lungs.

2. Expiration: It will be the opposite of what happens during inspiration: the diaphragm and the rib muscles will relax, and will diminish the thoracic cavity. This will provoke the passive departure of air.

Exchange of Gases
First, I don't know if I should first make the mixture of gases that have to breathe and adapt the respiratory system to them; or if I will do the reverse. I suppose that first I will make the mixture of gases and call them: "Air".

The exchange of gases between air and blood will take place through the thin walls of the alveoli and blood capillaries. The venous blood from the pulmonary artery will be released from the carbon dioxide, coming from the metabolism of all the cells of the body, and will take oxygen.

The oxygenated blood will return through the pulmonary vein to the heart that is pumped throughout the body.

Other systems I would create:

The Digestive System
As food is being digested, the blood flows through the intestinal capillaries and absorbs nutrients, such as

glucose (sugar), vitamins, and minerals. The blood distributes these nutrients to the body tissue.

Kidney and Urinary System
The wastes from the body tissue will be filtered through the blood as it flows by the kidneys. Then, the body will eliminate the wastes through the urine.

Temperature Control
To regulate the temperature, the organism will receive help from the blood flow that roamed through the different parts of the body, since body tissue will produce heat while they go through a breakdown of nutrients to turn them into energy, develop new body tissue, and eliminate waste.

Something is Missing...

What can I do so that it could defend itself from any strange attack?
I will make it a system of protection.
I will name it: Endocrine (Hormonal) System

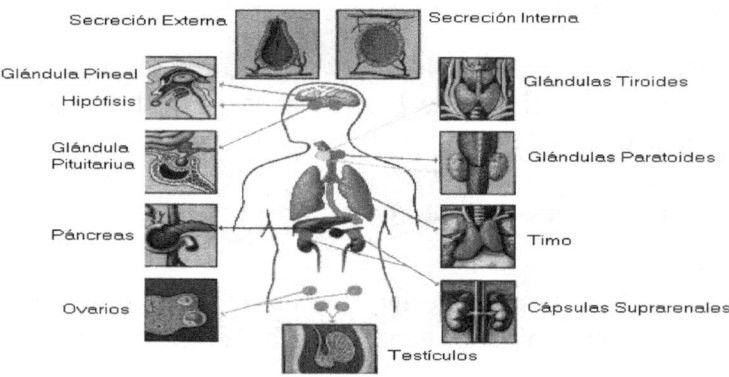

- The **Endocrine System** will be formed by a series of glands that release a type of substance called hormones; that is, it will be the system of the glands of internal secretion or endocrine glands.

- A hormone would be a chemical substance that would synthesize itself in a gland of internal secretion and would exert some kind of physiological effect over other cells until they reach the blood.

- **Hormones would act as chemical messengers** and would only exert their action on those cells that possess in their membranes the specific receptors (I would call these, target or white cells).

- The most important Endocrine glands that I would make would be Epiphysis or pineal, the hypothalamus, the pituitary gland, thyroid, parathyroid, the pancreas, the suprarenal glands, the ovaries, and the testes.

How will I go about so that it can reproduce?

Oh, the things that should be designed to enable reproduction and be able to have children!

Here are **some** details:

Fertilization (and the start of life), will happen when a sperm (of the 200 to 400 million available) penetrates the egg. The semen has to be the product of the secretion from the testicles, the epididymis (the gland that is attached to the testicle), the seminal vesicle, and the prostate.

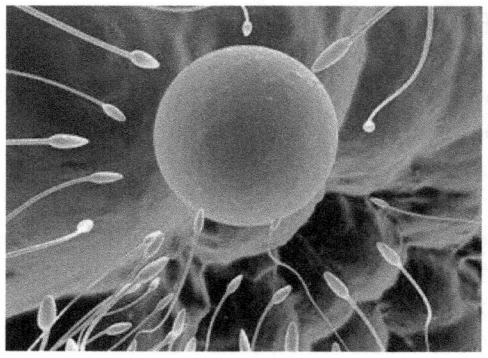

About sixty percent of its contents will be seminal liquid, thirty percent will be prostate fluid, and ten percent will be established by the sperm.

In addition, the seminal fluid will contain ascorbic acid (8 to 12 mg/dl), citric acid (350 to 600 mg/dl), fructose (200 to 400 mg/dl), and glicerilfosforilcoline (15 to 45 mg/dl). An ejaculation generally will produce from 1.5 to 5 cubic centimeters of semen.

The pH of this liquid will be from 6 to 9, its color a white opalescent, its liquefaction will be presented after 15 to 60 minutes and generally must be between **40 million to 250 million of sperm, for each cubic centimeter**, from which more than 50% must be alive, more than 50% of them must be in movement and more than 60% must have a normal aspect.

In order for the fertilization of the egg to produce, the semen must contain more than twenty million sperm per milliliter. If it is found to be below this figure, there would then be talk of male sterility or infertility…

...Do you grasp the idea?

What, how, and when did all those parts evolve to form that perfect and complicated process?
Not just in human beings, but also in all the animals, birds, etc?

The Bible says:

"As thou knowest not what is the way of the spirit, nor how the bones do grow in the womb of her that is with child: even so thou knowest not the works of God who maketh all." (Ecclesiastes 11:5)

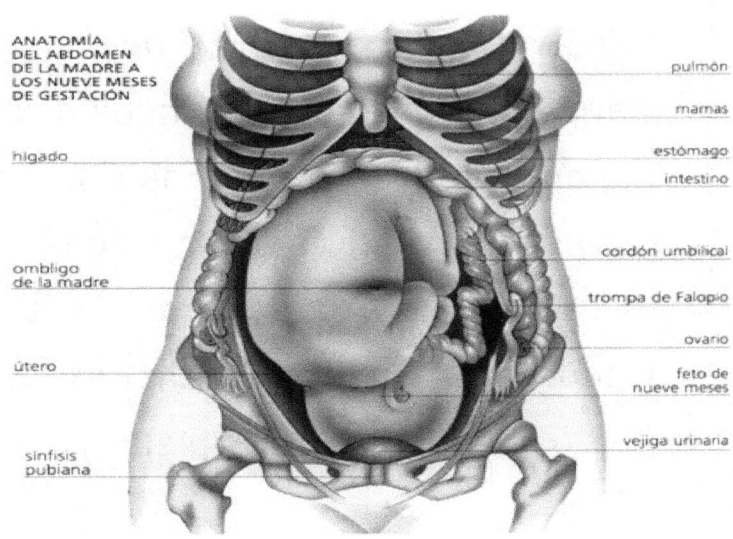

And what do we say about the design of the brain, the skin, the blood, the eyes, ears, smell, taste, touch...?

And they want us to believe that **"NOTHING"** is so smart and powerful that it could make all matter, space, and energy that is in the Universe be concentrated in a point smaller tan an atom; and give such precise and inviolable laws in everything that exists on Earth and in the heavens, that even in the super technological age and with the largest majority of scientists never concentrated on Earth, STILL we are learning the wonders of those things that exist, visible and invisible.

Instead of using the **logic** and **reason** for which we have been endowed; and instead of being sincere and grateful, we have preferred to encourage our unsatisfied pride because "we do not have all the answers"; and as a consequence of our denial, we live our lives without the sense and satisfaction that we should express.

Let us look at this small summary, about the theory of evolution [41]:

Many problems exist with the theory of evolution:
The Second Law of Thermodynamics does not allow disorder to produce order. The disorder or entropy of the universe never decreases but always increases with time.

How can it be that man, an extremely complex organism (order) could evolve from a simple organism such as bacteria (disorder)?

The probability that DNA will form randomly requires a lot more time than the age of the same universe. It would take more than 10^{75} years (the number 1 followed by 75 zeros).

Scientists cannot prove the origin of life of something without life.

The majority of mutations are harmful resulting in the extinction of the organism, although there are some that are neutral.

Evolution has no explanation for photosynthesis.

Evolution has no explanation for metamorphosis.

Evolution has a limited point of view since it only tries to explain the origin of species once life is already present on the planet and it ignores the total process of the creation of the universe.

Evolution has no explanation of how man acquired his state of conscious and intelligence.

The human eye and ear have an irreducible complexity to be a product of casualty. The eye without retina, or without optic nerves, or without cornea, or without iris, etc. is useless. It is impossible for this organ to "evolve" gradually.

Evolution has no explanation for the process of blood clotting.

When a person dies, the body generates endorphins to make death less traumatic. How could it be that evolution develops a process to make death easier, since this obviously does not offer any advantage for the survival of the organism?

How can it be that the butterfly passes through four stages of life (egg, caterpillar, cocoon, and butterfly) and only the butterfly is able to reproduce itself?

Evolution has no explanation for the altruism in animals and insects. For example, how can it be that an army of ants or bees (which cannot be reproduced) serve a queen (the only one who can reproduce)?

Evolution has no explanation of how the universe has been created with great precision for human life to exist.

This is an "anthropic principle".

In summary, it is quite obvious that a lot more faith is needed to believe in the theory of biological evolution than the one needed to believe that God is the creator of the universe.

■■■

This is true: If a person believes evolution has been real, have a **greater faith** than that of all the religious people.

As we have seen, the origin of the universe enters in the world of the **dogmas of faith.**

This is my suggestion:

- **Let us take out** the teaching of evolution from the public schools and universities, and the teaching of the creation.

- Let us use the education centers, to teach **science and not dogmas.**

- Pure Science is beautiful. Let us teach it!

If someone want to know more about origins, they need to be sent to the places where dogmas of faith are being taught (evolutionists will have to be considered as one more religion).

Now, if someone would ask me:

-You have studied;
How dare you believe in God?

Calmly, I can answer:

- Because I stopped believing

The Tale of Evolution!

■■

Speaking of the wonders that exist, we now will analyze something that, even though is very small at the same time is very complex: **The cell**

All cells have a common structure with three basic elements: the plasma membrane, the cytoplasm, and genetic material or DNA (deoxyribonucleic acid).

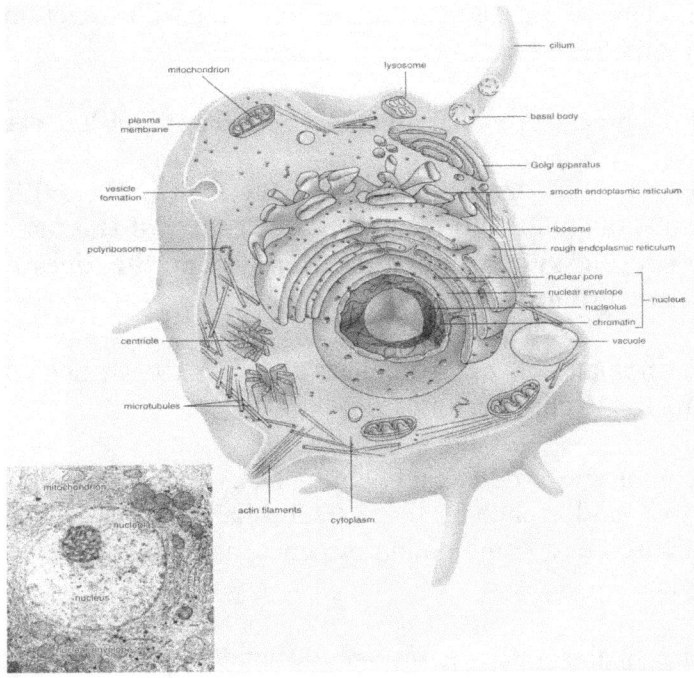

It was not until the end of the nineteenth century that was produced the cell theory, which says: "All living beings are consisting of one or more cells, i.e. the cell is the morphological unit of all living beings."

The cell contains all the information on the synthesis of its structure and the control of its function; and is able to transmit it to its descendants, i.e. the cell is the autonomous genetic unit of living beings.

The cell is capable of performing all the necessary processes to stay with life, i.e. the cell is the physiological unit of organisms.

Cells have the capacity to carry out three vital functions: nutrition, relationship, and reproduction.

All scientists know that **all cells come from another cell.**

"The human body contains close to 10 billion (10,000,000,000,000) of all types of cells: **of the brain, the nervous system, the muscular system, the digestive system, and others."**

Each human cell has a nucleus and inside each nucleus, there are 46 chromosomes.

Each chromosome has a chain of DNA that seems like a curved ladder called helix. Each chain of the DNA is composed of a combination of genes, as if each step were a gene.

These chains wrap themselves around the thousands of proteins that exist inside each cell and which are produced by the same cell. Without them, there would be no life.

All the information about the human being is written in the genes that make up the 23 pairs of chromosomes responsible for the specific function of each cell.

DNA is a chain that contains the code for all of our physical attributes as well as the instructions for all the functions of the body, including growth, development, and reproduction. In summary, genes are made up of DNA.

THE CASE OF THE GENOME

The dictionary defines **genome** as the ordering of genes that specify all the characters of an organism. Alternatively, it is all the genetic material of a living being.

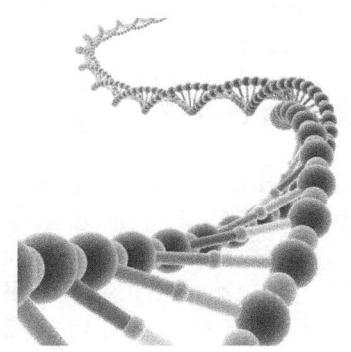

The genome is divided into chromosomes.

The chromosomes contain approximately 80,000 genes, which are responsible for the inheritance; and the genes are portions of deoxyribonucleic acid, or DNA.

In other words, the Genome is the complete set of hereditary instructions for the construction and maintenance of an organism that passes to the next generation.

It is the code that makes us be as we are.

In the year 2003, scientists complete the sequence of the Human genome.

The importance of fully knowing the genome is that all diseases have a genetic component, both the hereditary and the results from corporal responses to the environment.

Since the information contained in the genes has already been decoded, it will allow science to know, through genetic tests, which diseases may suffer a person throughout their life. Also with this knowledge, diseases that were once incurable may be able to be treated.

Do you know what the Bible says?
> God created man with the capacity to reason and understand; to learn and judge; and made him responsible for all the goods of the land.
> **(See Gen. 1: 26-28)**

[Notice that it does not say of the moon, or the sun, or the planets, etc.]

The Bible also tells us:

> "But God giveth it a body as it hath pleased him, and to every seed his own body. All flesh is not the same flesh: but there is one kind of flesh of men, another flesh of beasts, another of fishes, and another of birds". (1 Corinthians 15:38-39)

Let us be Good Stewards!
God expects that, as administrator of His goods, man will be faithful and not abuse of His privilege; respect the

design of what is created and not corrupt or alter the nature of what is under His control.

Is Man Delivering on His Word?

Let us analyze that question...

Around the year 1500, a false philosophical conviction, led an Italian man named Nicolas Machiavelli, to declare a phrase that has been the cause of great harm in society.

He said, "The end justifies the means".

Today, some unscrupulous people have decided to violate the nature of God's creation. Here are some cases...

Many other genomes have already been sequenced. Among them are the hen, mouse, and the chimpanzee.

Did you know that, the DNA of the mouse is closer to the human being than that of the chimpanzee?
[http://waste.ideal.es/genoma-raton.htm]

The facts published in the magazine "NATURE" in December 2002, tell us the following:

- The similarities between man and the mouse are, from a genetic point of view: equal in 99%.

- Only 300 human genomes are not present in rodents; only 300 genes from the mouse do not exist in our genome.

- According to British experts, "you could say that we are essentially mice without a tail, although we retain the genes that could lead to us developing the tail".

- However, only 96% DNA of the chimpanzee is similar to that of humans; and it is said that this number means "what keeps us away from these primates are 35 million different bases (the letters that make up the structure of DNA) and many chromosomal variations".

Could it be that somebody would now dare to say That we do not descend from the ape, but from the Mouse.
OH, NOOOOO!!!!!

What do you think some scientists are trying to do right now?

They started

the race

To

manipulate

DNA!

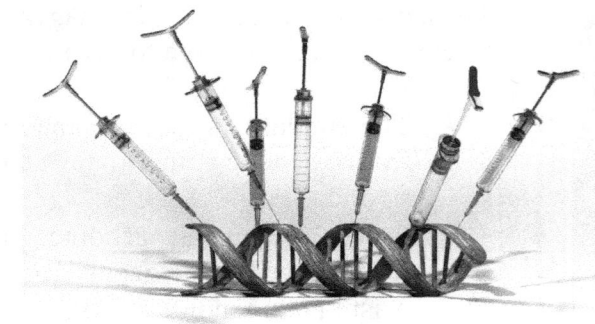

British scientists sought permission from authorities to create an embryo merging human Cells with cow eggs.

"Scientists plan to merge a cow egg with human DNA. The investigation seeks to study some of the most debilitating and incurable neurological diseases".

"For that purpose, researchers at King's College in London and the University of Newcastle requested a three years license at the Human Fertilization and Embryology Authority of Great Britain".

"The human-cattle embryo will be used to obtain stem cells and will only be allowed to develop for a few days".

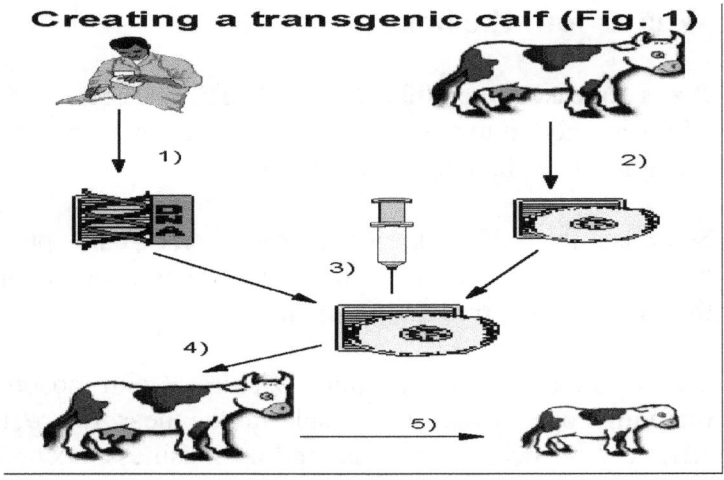

Creating a transgenic calf (Fig. 1)

The **procedure** used by the scientists to create a merging human cell embryo with cow eggs[43]:

1) A human gene responsible for producing a desired protein is isolated in a laboratory.

2) An animal is given hormonal treatment to produce a large number of **embryos**, and the embryos are collected from the oviduct.

3) The human gene is inserted into the fertilized egg via micro-injection. DNA of the pronucleus is injected into the fertilized embryo.

4) The transgenic embryo is placed in a surrogate host, which gives birth to the transgenic animal.

5) The offspring is then tested for the new gene.

What about the hen?

Press Release (Dec-2004) L.A. GAMEZ/IDEAL
Chickens and humans are more similar than previously thought: they share 60% of their genes [44]

Scientists from 12 countries- among them, Spain- present in the magazine "Nature", the sequenced chicken genome, the first decoded bird and farm animal.

This information, divided into 39 pairs of chromosomes- including one sexual-will result in the near future into advances in medical research and agribusiness, according to experts.

"Chickens and humans are, in some cases, infected by the same viruses, bacteria, and parasites", explained yesterday Jerry Dodgson, a microbiologist and molecular geneticist at the University of Michigan State (MSU) and one of the coordinators of the International Chicken Genome Sequencing Consortium.

Model Research

"We are more similar to the birds than what we thought. Around 60% of genes that encode proteins in the hen have their equivalent in man", emphasized yesterday Peer Bork, from the European Molecular Biology Laboratory, one of the 49 institutions that have participated in the project.

The chicken is a very important model for biomedical research because it is easy to maintain, it reproduces rapidly and it is easy to determine the different lineages by its physical characteristics.

Its genome is composed of 1,000 million pairs of chemical letters-one third of man-that are divided into 39 pairs of chromosomes, one of which determines the sex: the males have two Z chromosomes, while in the females that pair is formed by one Z and one W.

The chicken has between 20,000 and 23,000 genes, whose first studies have already given some surprises".

>>

Other related news:

British scientists create hens with human genes [45]

They are genetically modified and lay their eggs with useful proteins to manufacture anti-cancer drugs and other diseases. The experts belong to the Roslin Institute in Edinburgh, where the cloned sheep Dolly was created.

British scientists created various types of genetically modified chickens able to lay eggs that contain useful proteins to manufacture anti-cancer drugs and other diseases, informs today "The Sunday Times".

The experts at the Roslin Institute in Edinburgh (Scotland), where the cloned sheep Dolly was created, **have raised 500 laying hens** from a common species called ISA Brown, who's DNA manipulate the introduction to human genes producers of proteins. These human proteins are located in the egg whites, of which can be easily extracted for the development of pharmaceuticals, explains the newspaper.

One of the types of hen created by the scientists at Roslin produces interferon, an antiviral protein that is often used in the drugs against multiple sclerosis. Another type produces miR-24, which could be used on experimental medicine with potential to treat skin cancers and arthritis, according to the newspaper.

"This has the potential to become a very good way to produce specialized medicines", declared Karen Jervis, of the biotechnology company Viragen, who has collaborated on the Project.

"We have reared five generations of chickens and until now all of them continue to produce large concentrations of drugs", he added.
Source: EFE

What about the mice?
Current News: Source: Reuters 09-23-2005

Scientists implanted a human chromosome into mice [46]

London- Scientists have transplanted an almost completely human chromosome into mice making a big technical and medical step that could contribute a new comprehension of Down syndrome and other disorders.

In a published investigation in the **journal Science**, the researchers described how they withdrew the chromosomes from human cells. The chromosomes are at the core of the cell and contain the genes.

The human chromosome was mixed with embryonic stem cells from a mouse and a substance was added that made them both merge.

The stem cells that absorbed the chromosome 21 were then injected into the embryo of the mouse that was reintroduced to the mother.

The resulting mouse has a copy of the human chromosome.
>>

The subject of Cloning takes us further away...

Remember Dolly?
(The sheep that was cloned in 1996)

It was the first cloned animal from DNA resulting from an adult sheep instead of using the DNA from an embryo. To clone her, **277 attempts** were needed to produce this birth.

Well, now scientists are looking to clone **human beings.**

This perspective is that, obviously, has awakened a mixture of anxiety, fear, and fascination with the public's opinion.

> As it is said: The current citizen perceives the scientific advances with certain ambivalence: if he/she recognizes as positive the advancement of knowledge and welfare, and is equally aware that these can cause environmental problems and threaten the values and beliefs that are important for social cohesion.

As it is known, when a technique is used on a domestic or laboratory animal, it is only a matter of **time and money** when it can be applied to humans.

...It is only a matter of time and money!

I want to voice my opinion and join in the opinion of Mr. Horacio Ricciardelli, chairman of the Military Civic Movement CONDOR (Native Community of Regional Organizations), in the article I found on his web page: www.mov-condor.com.ar/cristianismoyvida/embrioneshibridos.htm, which I am transcribing below:

"Creation of 'cybrid' embryos"

At the beginning of September 2007, the British Human Fertilization and Embryology Authority (whose abbreviation is HFEA), published a study about the legalization of experiments that produce and use human "cybrid" embryos (a type of "hybrid" embryo).

How would those experiments function? The laboratories would take some animal eggs. After removing their nucleus, they would introduce in them the nucleus of a human cell, and then activate the egg in a way that it would develop as if it were an embryo.

To resort to the eggs of animals has various motivations. Among others, they avoid dangers that are given to women who participate as egg donors in these types of experiments. Also, it would greatly increase the availability of eggs for the experimentation, because there are fewer obstacles that exist when it is time to extract them from animals.

What would come about with these types of experiments? A living being would be obtained with human DNA in the nucleus and with other biological materials (cytoplasm, mitochondria) non-human.

This living being has been called "cytoplasmic hybrid embryo" (also called "cybrid embryo"), and it would contain human DNA and animal DNA (present in the mitochondria).

The human DNA would be more than 99% of the DNA total, so it is supposed that this embryo would be practically a human embryo, even though it is on this point that there are some doubts in the scientific world, as will be mentioned later on.

The Project is not entirely new, because there have been some experiments of this type in China and in the United States. In addition, we must take into account, that in Great Britain the law allows you to produce human embryos for research (i.e., intended to be destroyed in an experiment).

What is now being proposed is to authorize the creation and use of cytoplasmic hybrid embryos.

It should be made clear that we are not before a clone, but before a special hybrid, in which initially the cytoplasm is not human and the nucleus is human. Therefore, it would not be, a hybrid in strict sense, it would be possible through the fertilization of an animal egg with a human sperm (or a human egg with the sperm of an animal).

Therefore, we do not speak about normal hybrid embryos (in which case this type of hybridization would be possible to achieve), but of the eventual creation of "cytoplasmic hybrid embryos".

The questions before this proposal are many. The first, the most important, is: what is obtained when making a cytoplasmic hybrid embryo? Is it really an embryo? Is it a human embryo?

For some, the answer would be yes. But others have serious doubts: would it be about a new animal species on earth, halfway between human and non-human?

Would it not be simply just a handful of disorganized cells, and therefore, it would not become a real embryo?

At the hour of delivering an ethical judgment, we must take into account the various alternatives.

If the result of the experiment was a human embryo, it would deserve the respect of every human being: it would be unfair to produce and create it, to then destroy it, as it is already unfair to create and destroy human embryos using human eggs.

For some the hybridization will produce a "special" or rare human embryo, this does not take away its value, its dignity: every human being deserves to be respected and accepted, defended, and treated simply for being who they are, by their human condition, which makes them worthy of a fair treatment and protection before any type of aggression by others.

As for those who have serious doubts about the human condition of the cytoplasmic hybrid embryo, ethics tells us that nor in this case would it be lawful to undertake these experiments, while still being in a state of doubt.

It will never be correct to use and destroy a living being produced in a laboratory on the question of whether it is nor is not a human individual.

In case there is any doubt, we cannot work with biological realities which could have the value of every human being.

In the same way, these types of experiments should be prohibited as long as the question remains of whether it would produce human beings, even if they were "special" human beings: having a special or "rare" difference should not become a reason to treat a human being as a laboratory animal.

The second question revolves around the ending for these types of experiments.

They tell us, the creation of hybrid embryos would produce stem cells with the same DNA of people with diseases such as Alzheimer's, Parkinson's, or other ones similar to these, to see how these such cells would develop and thus study the various ways of preventing or mitigating these diseases.

The ethical judgment on the alleged therapeutic order of these experiments depends on the response to the first question: what is obtained when making in the laboratory in introducing human DNA in an animal egg where the nucleus has been removed?

If it is about human embryos, they can never be used, not even for the progress of medicine, as it goes against ethics and against justice.

If there is no clear answer, it would not be lawful to use these embryos while still in doubt.

We should remember that a good ending cannot justify a bad means.

Discovering new treatments for extremely painful diseases does not make an experiment ethically good that can go against the dignity and against the life of embryos that are human or on whether there is a minimum doubt about its possible human condition.

The scientific research on hybrid embryos is, therefore, ethically reprehensible.

The doctors, scientists, and the whole society show their love of justice and respect for fundamental ethical principles if they reject and achieve an experimentation so full of doubts and so contrary to the respect that every human being deserves, albeit a small and helpless embryo.>>

We know that the subject of cloning is not entirely new. I present a summary of the major investigations that have been made on cloning[47]:

1952: Robert Briggs and Thomas King-University of Pennsylvania (USA)-the obtaining of frogs from cells of an embryo.

1967: John Gurdon-The cloning of Xenopus laevis (African clawed frog) on the basis of using intestinal cells from adult animals. The animals died without becoming adults.

1980: Allegheny University of the Health Sciences-San Luis (USA)-The cloning of tadpoles from red blood cells. This is the same as in the previous case; the animals would die without reaching the adult stage.

1981: The cloning of mice becomes available, but the animals would die in an embryonic state with severe malformations.

1985: Steen Willadsen-Institute of Animal Physiology, Cambridge-the obtaining of sheep from embryonic cells.

1986: Neal First-University of Madison (USA)-cloning of the first cow from an embryonic cell of six days.

1995: Sir Ian Wilmut and Keith Campbell-Roslin Institute at Edinburgh (Scotland). Birth of the sheep Megan and Morag, cloned from fetal cells.

1996: Sir Ian Wilmut and Keith Campbell-Roslin Institute at Edinburgh (Scotland)*-Birth of Dolly the sheep, cloned from the breast cells of an adult sheep.

1997: Roslin Institute at Edinburgh (Scotland)-Birth of Polly the sheep, cloned from fetal cells. She is also transgenic.

1997: University of Massachusetts (USA)-cloning of veal calves from embryonic cells of a connective tissue.

1997: University of Hawaii (USA)-cloning of dozens of mice from follicular cells.

1998: France-birth of the cow Marguerite, cloned from fetal muscle cells.

1998: Ishikawa (Japan)-the obtaining of calves from the intestinal cells of a cow.

1999: Virginia (USA)-birth of five cloned pigs, obtained from the same procedure as Dolly.

(*) The great novelty that was achieved in the year 1996 is that Dolly the sheep was the first cloned mammal from the cells of an adult animal. It had already been achieved before starting from embryonic or fetal cells. >>>

But...

Let us remember that God expects for man, as administrator of His goods, to be faithful.

Part of being faithful consists of not abusing of his privilege, **respecting** the design of what is created, and not to **corrupt or alter the nature** of that which is under his dominion.

What is happening then?

We have seen that **evolution**, which says that from something smaller than an atom everything that exists was formed, it **does not serve** to give a reasonable explanation of the countless variety of fauna and flora (animals, vegetables, mosquitoes, and flowers...) minerals...(Not to mention our real and own existence as rational and thinking human beings).

Then, we ask ourselves:
Why do so many people tenaciously cling on to a belief that has so many shortcomings?

What is it that has changed in our world?

A few generations ago in some countries and communities, it was prohibited to teach the theory of evolution. In general, the Bible was accepted as a true and trustworthy story in our origins.

However, today, very different concepts dominate. The Bible is practically banned in schools, and a serious study from the biblical point of view about the creation of the universe-and the origin of man-is prohibited.

At the same time, on some occasions the critical analysis over the theory of evolution is dropped sharply in the academic and scientific circles.

Why is it that in schools and universities it is being taught that evolution "is a proven fact" when this is a terrible **lie**, since **no science** (not even the genetic, nor the paleontology, nor astronomy, nor geology, nor biology, nor zoology, nor anthropology, nor physics, nor chemistry, etc.) has NEVER been able to verify any evolutionary theory?

As we will see, what the Apostle Paul commented on the philosophers of his time, can also be applied for todays.

> "Because that which may be known of God is manifest in them; for God hath shewed it unto them. For the invisible things of him from the creation of the world are clearly seen, being understood by the things that are made, even his eternal power and Godhead; so that they are without excuse".

> "Because that, when they knew God, they glorified him not as God, neither were thankful; but became vain in their imaginations, and their foolish heart was darkened. **Professing themselves to be wise, they became fools...**" (Romans 1:19-22)

Religion is not free of guilt

According to the words of British physicist Alan Hayward:[48]

"When the first fathers of the church claimed that the world was flat, they believed they were defending what the Bible said.

But what was happening in reality was that they were **defending their own erroneous interpretations of the Bible.** In the long run, what they achieved with this behavior was giving people the impression that in the quest for knowledge, Christianity was opposed to the scientific method".

It should be noted that the first battles between scientists and the Bible took place due to the erroneous interpretations of the Bible; NOT to what really says in the Word of God.

As it is said to us by Starr-Taggart in their compendium titled, "Biology: Unity & Diversity of Life"[49]:

"From time to time, scientists dispute controversies when they explain something that was believed to be beyond a natural explanation or that it belonged to the supernatural.

This is the case often when moral codes of society are found interwoven with religious narrations. To explore a lasting opinion from the natural world from a scientific point of view could be misinterpreted just like questioning morality, although both concepts are different".

"For example, several centuries ago, in Europe, Nicolaus Copernicus studied the planets and came to the conclusion that the earth revolved around the Sun.

In reality, this now seems obvious to us, but back then it was considered a heresy".

"In that time the prevailing belief was that God had created the Earth (and by extension to humans) and had placed it as the unchanging center of the Universe".

Subsequently, a respected scientist, Galileo Galilei, studied the model of the solar system of Copernicus, and considered that it was good and he said so. However, he was forced to retract publicly on his knees, and had to say that the Earth was the center of all things".

It is also my view, according to what the author declares, that "this does not mean that scientists who formulate questions are less moral, do not respect the law, are less sensitive, or do not care about the fate of other people;
It simply means that their work is governed by an additional norm: *the external world and not the inner conviction should constitute the field of proof of scientific beliefs".*

However, we should not leave aside logic when we do research.

...We see that the Darwinian Theory has been widely accepted, most of all, as **a cry of protest**;

However,

As a result, a great deal of moral and social harm has been made and it has had a huge and terrible effect on thousands of people.

It so happens that, the spirit of the antichrist, which has been in the world since the time of the apostles of the Lord and it opposes everything that has to do with God and Christ, has taken a great advantage of this situation.

With the power of **deception** that it has, it has shown the bad to be good, and the good to be bad...And it is provoking that millions of people to not believe in God and they call themselves atheists.

We must understand that the struggle being waged

For the souls of human beings,

Is Spiritual!

The Bible warns us:

"Be sober, be vigilant; because your adversary the devil, as a roaring lion, walketh about, seeking whom he may devour; whom resist stedfast in the faith" (1 Peter 5:8-9)

The Bible also tells us:

"Finally, my brethren, be strong in the Lord, and in the power of his might. Put on the whole armour of God, that ye may be able to stand against the wiles of the devil.

For we wrestle **not against flesh and blood, but against** principalities, against powers, against the rulers of the darkness of this world, against spiritual wickedness in high places." (Ephesians 6:10-12)

Darwin's theory led him and the education system as well, to reject the existence of God and to rule out the Bible in the classrooms and in all of society.

It is not a coincidence that Karl Marx, the father of communism, asked Darwin if he could dedicate his masterpiece "Capital" or if Darwin was willing to write the prologue; because Karl Marx believed that Darwin had given him the scientific bases for communism.

It is said that Darwin discreetly declined the offer...[50]

Later, Adolf Hitler did in fact apply to the human race the Darwinian concept of "the survival of the fittest".

During the Second World War, the Nazis sterilized more than two million people and began to systematically exterminate those whom Hitler considered to be inferior.

The Nazis justified their acts by saying that they were doing a favor for humanity, because they were carrying out **"a genetic purification" to improve the human race**[51]

No words…

By the way,
Did you know that **evolution** is the **only one** philosophy that can be **used to justify**:

- The hundreds of million children ripped apart and destroyed before they saw the light of the day, by means of the abominable WORLD-WIDE INFANTICIDE, called Abortion

- The rampant immorality and sexual disorder that the Pornography has carried out

- The scorn to the life of other human beings, under the image of Racism or Nazism

- Other great disorders of conduct that today we see in our modern society

Should we be silent or speak?

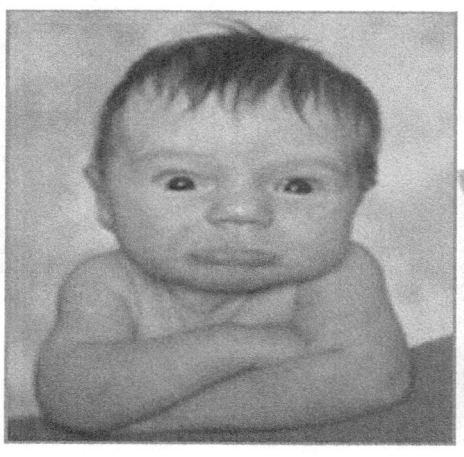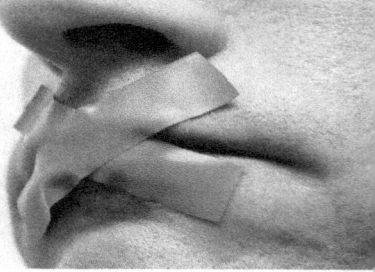

The Bible tells us this:

"I charge thee therefore before God, and the Lord Jesus Christ, who shall judge the quick and the dead at his appearing and his kingdom. Preach the word; be instant in season, out of season; reprove, rebuke, exhort with all long-suffering and doctrine.

For the time will come when they will not endure sound doctrine; but after their own lusts shall they heap to themselves teachers, having itching ears. And they shall turn away their ears from the truth, and shall be turned unto fables. But watch thou in all things, endure afflictions, do the work of an evangelist, make full proof of thy ministry".

(2 Timothy 4: 1-5)

Is it true that we have alternatives and that there are 2 roads?

Jesus said:

"Enter ye in at the strait gate: for wide is the gate, and broad is the way, that leadeth to destruction, and many there be which go in thereat: Because strait is the gate, and narrow is the way, which leadeth unto life, and few there be that find it."

(Matthew 7:13-14)

The two roads are real.
It all depends on what is the way you are touring when your time comes to leave this world. Do not let yourself be fooled. It is **YOUR** own soul that is in danger.

The future of the Soul.

In the book of **1 Thessalonians, chapter 5-verse 23,** the Bible tells us that all human beings have:

- Spirit, Soul, and Body

The Bible also tells us in the book of **Ecclesiastes**, that when death arrives, and we have to leave this world:

> "…Then shall the dust [*body*] return to the earth as it was: and the *spirit* shall return unto God who gave it." (Ecclesiastes 12:7)

What of the **soul**?

It will go to one of two places: Heaven or Hell.

Out of everything you own, the soul is the most valuable.

Jesus also said:
> *"For what is a man profited, if he shall gain the whole world, and lose his own soul? Or what shall a man give in exchange for his soul?"* **(Matthew 16:26)**

> *"And fear not them which kill the body, but are not able to kill the soul: but rather fear him which is able to destroy both soul and body in hell."* **(Matthew 10:28)**

Levels of Responsibility:

Do you want to know what you should do?

Recognize that
You also

Will have to appear before the court of God...

The Bible tells us:
> "Look unto me, and be ye saved, all the ends of the earth: for I am God, and there is none else." (Isaiah 45:22)

> "... And walk in the ways of thine heart, and in the sight of thine eyes: but know thou that for all these things God will bring thee into judgment."

> "Let us hear the conclusion of the whole matter. Fear God, and keep his commandments: for this is the whole duty of man".

> For God shall bring every work into judgment, with every secret thing, whether it be good, or whether it be evil." (Ecclesiastes 11:9; 12: 13-14)

(See APPENDIX #4, for more information)

In addition, seek to know more about God and of His plan for your life... Read and apply the Bible. Talk to others of what you have seen and heard. You will see surprising results!

Then spoke Jesus again unto them, saying, "I am the light of the world: he that followeth me shall not walk in darkness, but shall have the light of life." **(John 8:12)**

Another thing that you should do is:

Teach your children.

The Bible says:

> "Train up a child in the way he should go: and when he is old, he will not depart from it." (Proverbs 22:6)

Make yourself responsible in teaching your children! Do not let any system be responsible for something as vital as the future of your own children.

Do your part! (The possible)...

You will see how **God will do His** (The impossible)

(See in APPENDIX #5, some suggestions on how we can talk to our children about this subject).

Good news that encourage and give us hope:

1. MATTERS OF LIFE AND DEATH
Vote will protect unborn beginning at conception[52]
'Victory serves as example to other nations'

Posted: April 24, 2009 By Bob Unruh
© 2009 WorldNetDaily

Lawmakers working on a new constitution for the Dominican Republic have voted overwhelmingly to protect life, specifying in the document that "the right to life is inviolable from conception until death."

The vote yesterday was 167-32 in the national legislature, which was responding to pressure from international pro-abortion groups seeking to expand their business operations into the Caribbean island nation.

2. Lawmakers declare fetuses to be people, too[53]
States vote on measures that extend full 'personhood' rights to pre-born

Posted: February 28, 2009 By Drew Zahn
© 2009 WorldNetDaily

Legislative bodies in two states voted this month to define the beginning of human life – and human rights – at conception.

On Feb. 17, North Dakota's House of Representatives voted 51-41 to approve a bill that declares "any organism with the genome of homo sapiens" – even one not yet born – is a person protected by rights under the state's constitution.

Yesterday, the Montana Senate voted 26-24 to approve S.B. 406, a constitutional Personhood Amendment that states, "All persons are born free and have certain inalienable rights. ...

Person means a human being at all stages of human development of life, including the state of fertilization or conception, regardless of age, health, level of functioning or condition of dependency."

Both bills, which have major implications for abortion should the states grant unborn babies the full status of "persons" under the law, now await approval by their respective opposite legislative houses: the North Dakota Senate and the Montana House of Representatives.

Should S.B. 406 pass in Montana's Legislature, it would then be sent to the voters of the state, who with a simple majority could make it part of the state's constitution.

3. Darwin's theories are prohibited in Serbia [54]

Events September 17, 2004

<<BELGRADE, (Reuters).-The Serbian Education Minister, Ljiljana Colic, ordered the Serbian schools to stop teaching children the theory of evolution during this year and to resume this subject in the future, unless they introduced creationism.

The decision surprised the educators and the editors of textbooks in the ex-communist state, where religion had remained outside of education and politics and very recently was allowed entry into the classrooms.

Darwinism "is a theory as dogmatic as the one which says God created the first man", said Colic to the daily newspaper, Glas Javnosti.

Colic, an Orthodox Christian, ordered that the theory of evolution be discarded from the course of biology of this year for the 14 and 15 year old adolescents in their last grade of middle school.

Starting next year, both creationism and the theory of evolution will be taught, she added.

"Creationism teaches that a supernatural being created man and the universe."

The majority of scientists consider "the creation" as a religious dogma and not a science.

"Both theories exist in parallel and legitimately in the rest of the world", said Colic.

"Evolutionism, which says that man descends from the ape and the one that says Almighty God created man and the rest of the world", she said. .>>

God willing that in all parts of the world, courageous people will rise, that dares to resist and not believe

"The Tale of Evolution"

So that all, one day we can Unite ourselves to the voices that, without ceasing, declare in Heaven:

"Thou art worthy, O Lord, to receive glory and honour and power: for thou hast created all things, and for thy pleasure they are and were created."

"And when those beasts give glory and honour and thanks to him that sat on the throne, who liveth for ever and ever, The four and twenty elders fall down before him that sat on the throne, and worship him that liveth for ever and ever, and cast their crowns before the throne, saying, Holy, holy, holy, LORD God Almighty, which was, and is, and is to come."

(Revelation 4: 8-11)

To the King of the centuries, immortal, invisible, to the only and wise God, be the honor and glory forever and ever, Amen.

Dear reader: The God who created the Heavens and the earth bless you and keep you whole!

Julio A. Rodríguez, IQ

APPENDIX #1

A. The singularity of the Bible

Why the Bible, and not any other book?[55]

Because the Bible is the most special book that exists. It is the book of books.

- It really is a <u>great Encyclopedia</u>, composed of 66 books.

- The Bible took up to **1,600 years** to be written

- It was written in **three languages** (Hebrew, Aramaic, and Greek) by about **40 authors**, and it is all **internally coherent.**

- It was written in three continents: Africa, Asia, and Europe

- It was written by very diverse people: prophets, priests, cupbearers, fishermen, etc.

- Until 1997, the Bible has been translated entirely or partially to almost 2,200 languages and dialects, putting the Scriptures to the extent of more than 90% of the world's population.

B. Reliability of the Biblical documents:

- The Bible is 98.5% literally pure. This means that throughout the whole process of repeatedly copying the Bible throughout the centuries, there is only about 1.5% doubt about its text.

- There is no other work that exists at all between the writings of old centuries that even comes close to the precision and accuracy of transmission, which are found in the biblical documents.

- The 1.5% doubt that exists about the text does not affect in any way the doctrine. These "mistakes" are called textual variants and mainly consist in modifications of words and spelling.

- The Old Testament does not have so many manuscripts that support it as does the New Testament, but it is still extremely reliable.

- The Septuagint, a translation of the O.T. Hebrew to the Greek carried out between the III and II centuries before Christ; testify to the reliability and consistency of the O.T. when compared with the existing Hebrew manuscripts.

- The Dead Sea Scrolls, discovered in 1947, also attest to the reliability of the O.T. manuscripts.

- The Dead Sea Scrolls are ancient documents that were hidden in caves in the Judean desert close to 2000 years. Among them were complete copies or fragments of almost all the books of the O.T. Among them, there was a complete copy of the book of Isaiah.

- Before discovering the Dead Sea Scrolls, the most ancient existing manuscript from the O.T. Hebrew dated from approximately 900 years after Christ (A.D.) and constituted of the so-called Masoretic (from the Hebrew Massorah tradition).

- The Scrolls contained biblical manuscripts that were 1000 years more ancient. The comparison between the two groups of manuscripts showed an accuracy of precision despite repeated copies, which many critics saw that they were forced to remain silent.

- The New Testament has the support of more than 5000 Greek manuscripts currently in existence, with 20,000 more in other languages (ancient translations in Syriac, Latin, and Copyto, etc.).

Part of the handwritten evidence

Part of the evidence includes copied manuscripts less than a century after writing the originals. The textual variation in the N.T. is inferior to 1%.

Estimated dates of production of the N.T. documents

- The letters of Paul, 48-66 A.D.

- The Gospel of Matthew, 70-80 A.D.

- The Gospel of Mark, 50-65 A.D.

- The Gospel of Luke and Book of Acts, 60-65 A.D.

- The Gospel of John, 80-100 A.D.

- Revelation, 96 A.D.

Some of the main existing manuscripts from the N.T. are:

- The John Rylands Papyrus, written about 130, the oldest know fragment of the N.T.

- Bodmer II Papyrus (between 150 and 200).

- The Chester Beatty Papyrus (200) contains a great part of the N.T.

- Codex Vaticanus (325-350), contains almost all the Bible.

- Codex Sinaiticus (350) contains almost all of the N.T. and more than half of the O.T. (Greek version).

No other ancient work can boast of having copies so close to its time of writing. For the Bible, the difference is 50 years. By comparison, for Plato or Aristotle, the difference is measured in **centuries.**

The probabilities that Jesus fulfilled 48 of the main 61 prophecies concerning to Him are 1 in 10ee157; this represents one divided to 1 followed by 157 zeros.

However, Jesus fulfilled them!

C. The Bible is also: UNIQUE IN ITS SURVIVAL[56]

Survival throughout Persecution

As no other book, the Bible has endured the malicious attacks of its enemies. Many have tried to burn her, to prohibit her, and "put her outside the law, since the days of the Roman emperors to the present, in countries dominated by communism."

Sidney Collett, in "All about the Bible", says: "Voltaire the prominent incredulous Frenchman that died in 1778, said that one hundred years after his time, Christianity would be erased from existence and would pass into history.

But what happened? Voltaire has passed into history; while the circulation of the Bible continues growing in almost all parts of the world, making blessings wherever it goes.

For example, the English Cathedral in Zanzibar is constructed on the grounds of a former slaves market; and the table of communion is on the same place where at one time there was a whipping post! The world abounds with similar instances…

As someone said with much certainty, "Attempting to retain the circulation of the Bible would be the same as putting our shoulder against the burning wheel of the sun and trying to detain its flaming course."

In regards to the boasting of Voltaire to the extinction of Christianity and of the Bible in 100 years, Geisler and Nix point out that "only fifty years after Voltaire's death the Bible Society of Geneva used the same press and house of his to produce tons of Bibles." WHAT A HISTORIC IRONY!

In the year 303 AD, Diocletian issued an edict (Cambridge History of the Bible, Cambridge University Press, 1963), to destroy the Christians and their sacred book:

"…It was enacted everywhere an imperial letter, ordering that churches be demolished and the Scriptures be destroyed by fire, and proclaiming that those who held senior posts would lose all civil rights, while those who were in their homes who persisted in their profession of Christianity, would be deprived of their freedom."

The historic irony of the previous edict to destroy the Bible is that Constantine, the emperor that followed Diocletian 25 years later, commissioned Eusebius to prepare fifty copies of the Scripture at government expense.

The Bible is unique in its survival. This does not prove that the Bible is true. No, but it proves that it remains alone among books. A student who walks in search of the truth should consider a book that has the previous unique qualities.

APPENDIX #2

THE METHOD OF RADIOCARBON DATING
By: Robert L. Whitelaw

The method of radiocarbon dating was first proposed and put in details by Willard F. Libby, who also received a well-deserved Nobel Prize in 1960.

Making unfortunate measurements on all kinds of living matter throughout the world, Dr. Libby was able to demonstrate that all living cells possess the same specific radioactivity because of the presence of approximately 767 atoms of Carbon-14 per thousand million atoms of Carbon-12.

While the cell lives, this proportion is maintained by means of a continuous cycle established between the living matter and the carbon dioxide in the air and sea, which is known by the name of "deposit of exchange of carbon."

Then he demonstrated, by means of atmospheric measurements in various latitudes and altitudes, that the velocity at which Carbon-14 will be replenished in this deposit by the action of cosmic rays are *reasonably close* to the velocity at which it disintegrates in living matter.

Then, he *assumed* that these two velocities are essentially the same, and that so have been for many years.

In this way was "born" the method of radiocarbon dating that has been used by scientists since then, a time period of 20 years [in 1970, in which this article was published; N. of T.].

The validity of the two previous assumptions will be considered later on. Let us assume that they are correct, and let's see how simple and safe this method is.

It is simple to calculate the number of years since the living matter of the specimen died until now, measuring the radioactivity that it presents.

After 5,570 years,[57] the clicks per minute in Geiger counters will be half of which would have been registered at the time of death; after 11,140 years the counting would descend to the quarter part; after 22,280 years, we would count a one-sixteenth part; and so would keep diminishing.

The only thing that is needed is a pure sample without mixtures of other dead or alive materials over the years, in addition to the assumption that the radioactivity that the specimen possesses in the time of his death was the same as that which the living matter exhibits in actuality; or 16.0 disintegrations per minute and gram[58] of total carbon (dpm/g).

Among the first specimens that Libby and his collaborators dated were some tree rings and age relics "of known ages" from ancient Egypt. The tally was quite satisfactory.

In 1952, the method was published in book form[59], together with 200 datings of archaeological and geological specimens gathered from 30 very separate localities.

A second edition was published[60] in 1955, and included a special appendix at the end of the greater part of the chapters in the 1965 reprinting of the second edition.

Once the new radiocarbon clock was established, university scientists and research centers around the world joined together to study this new field of investigation, setting up their own laboratories of dating. Towards the end of 1968, there were almost 100 occupied laboratories, as it is shown in Table 1.

The C14 was recognized as a valuable tool to identify the age of ancient cultural deposits and artifacts, as well as for the dating of pollen, trees, and buried vegetation, as well as bones and all types of relics from the past.

At the same time, all those involved in the method recognized that the method could give computable ages until only 50,000 years before our era, since the radioactivity of any previous object would be barely detectable.

With all certainty, would remain out of the scene, the possibility of fossil dating, petrified material, carbon, petroleum or the bones of prehistoric men or animals.

Using evolutionist premises, scientists have assigned these materials, ages much higher than 100,000 years; and many of them, within the scope of millions of years.

In short, it was only considered susceptible of dating, the material preceding of Pleistocene superior and Holocene. It was unthinkable to obtain a dating of the Tertiary strata, and was expected with certainty that a large number of specimens would give "infinite" ages, that is, too old to measure.

What has the results been? In a single word: Amazing! It has been Amazing for any researcher with evolutionist presumptions. But it is even more amazing, when they are being compared with the biblical record, as we will see.

LIST OF TEN AMAZING FACTS

Beginning with the first group of 200 dating, published by Libby in the first edition, the list has now grown, and towards the end of 1969, includes some 15,000 dating of independent specimens of all kinds gathered from all parts of the world by the ninety-one laboratories listed in Table 1.

(The wide distribution of these specimens by category and geography is given in Tables 2 and 3 of the study).

All of these datings were published, until the year 1958, in *Science*; and still today in the annual magazine *Radiocarbon*, with extensive details of the tested material and the location of origin of each specimen.

Recapitulating, the registry of the dating with radiocarbon is so large and broad, as far as times, localities, and tested materials, that no informed scientist, nor any historian, educator, nor editor; it doesn't matter how attached they are to the evolutionist premises, can excuse themselves of examining and of not stopping to consider its profound implications.

After considering these datings, and after verifying the descriptive material, there can be detected at least, **ten amazing facts:**

1.   Practically, every specimen of material that lived in the past has been dated within the past 50,000 years. Very few are dated until 60,000 and only 3 -three among 15,000- have the claim to be of "infinite" age. These three are some eggs of Megapodes, proceeding from a cave in the Philippine Islands.

> (*Note:* To fully appreciate the significance that this has, we should emphasize that if the Lyelliane geology and the evolutionary scale of "time" were lives, if living matter has been accumulating and dying upon the land over alleged periods of time, then a random world sampling of buried organic material as the

present, should show 20,000 non-datable specimens for every one that is datable!)

We suppose the fact that many researchers were making searches in ancient cultures, specifically India, Maya, Babylon, etc. We found that all of them are dated within 50,000 years *up to the maximum depth of all deposits*. The vast majority of samples are related to vegetation, pollen, peat bogs, buried trees, clay fossils, samples of ocean bottom, buried bones, and cultural fields of charcoal; most of which should have given "infinite" age.

However, they present them as a measureable radiocarbon activity!

2. Samples identified by the investigator in strata such as Pleistocene, Pliocene, and Eocene (or, 50 millions of years old for an evolutionist!), and the majority of findings identified as Paleolithic, appear aged less than 40,000 years.

3. Even the carbon, petroleum, and natural gas and lignite are dated within the last 50,000 years. However, the accepted carboniferous period that allegedly produced these materials was about 100,000,000!

4. Of the oldest ages, the major part belongs to buried vegetation of all kinds.

5. Some 22 dated specimens are identified as "fossils", semi-petrified materials, or fossilized layered materials.

6. Many datings come from extinct flora and fauna, which until now were attributed to the inferior and middle Pleistocene, such as the mammoth, melodeon, saber-tiger, etc. Almost all have been dated between 10,000 and 30,000 years.

7. Many remnants of "prehistoric" men and artifacts are dated within the past 30,000 years, including some cases as The Neanderthal Man, Broken Hill Man, Florisbad Man, Heidelberg, of Keilor and Hotu. In addition, there are doubts being cast on the dating between two and four million ages of antiquity[61] attributed by Leakey *et al.* to forms such as *Zinjanthropus* of Olduvai and the *Australopithecus* from the Omo Valley! [62],[63],[64]

8. The deposits from the ocean floors and samples of 14 meters of depth from the ocean bottom, which is assumed that it contains the detritus of more primitive life forms, are dated within the past 40,000 years.

9. The ancient artifacts dated by archaeology (in Egypt, Syria, Iran, etc.) show in general that the radiocarbon datelines are 500,000. But today is recognized the tendency of exaggeration by the ancient historians. (Beroso, Maneto, etc., N. of T.)

10. Most ancient ages of the human culture is found in the Middle East, while the most ancient "human" datelines of the Western Hemisphere are noticeably more recent. To substantiate the dramatic findings of the previous numbers (3), (5), (6), and (7), Table 4 *(in the original study)* gives the list of 75 typical datelines of more than 220 that have been found in these categories to date.

These facts have already disrupted some specialists in evolutionary geology and paleontology, as evidenced by a typical assertion in Science (October, 1956): "Because of the radiocarbon dating, all previous interpretations from the history of the Pleistocene, its depth and position in the geologic column, must be reviewed" (p.669).

However, most disturbing are still the facts that arise from a more careful analysis made throughout this great collection of data.

Here we have before us, gathered from all parts of the globe and covering almost all forms of life already dead, a sufficient number of datelines of deaths to learn something, thanks to its distribution. If they are distributed by age, by location, and by type, in accordance to the indications of some old historical record, it should not be difficult to confirm or refute such registry.

Let us consider as an example a chronology based on the Bible *(see Table 5 of the original study)*. The Bible describes a creation only a few 7,000 years ago, then some 2,000 years later a global catastrophe that almost completely extinguished humanity, animals, and the birds from the face of the earth.

Now that we have a wider sampling of datelines of death that take us to the most primitive principles of man, surely we can reject such a strange document in a definite manner!
Or, is there any possibility that it may have corroboration?

171

APÉNDICE #3

Biography of Charles Darwin[65]

The biography of Charles Darwin begins with his birth on February 12, 1809 in Shrewsbury, England.

Darwin was a British Naturalist who became famous for his theories of **evolution and natural selection**.

Darwin believed that all life on Earth evolved (it gradually developed) by millions of years from a few common ancestors

From 1831 to 1836, Darwin collaborated as naturalist aboard the **HMS Beagle** on a British scientific expedition around the world.

In South America, Charles Darwin found fossils of extinct animals that were similar to the modern species.

In the Galapagos Islands of the Pacific Ocean (Northwest of South America), Darwin noted **many variations** between the plants of Galapagos and animals of the same general type to those found in South America.

The expedition visited various places around the world, and Charles Darwin studied the plants and animals in all the places visited; collecting species for subsequent studies.

According to the biography of Charles Darwin, we can note that since his return to London he conducted a comprehensive work of investigation on his notes and specimens.

From his study, various related theories were born:

- Evolution did happen.

- The evolutionary change was gradual, requiring thousands to millions of years.

- The main mechanism for evolution was a process called **Natural Selection** and

- The millions of species that now live emerged from a single form of original life through a process called **specialization**.

The theory of Darwin of evolutionary selection sustains that the variation between species occur randomly and the survival or extinction of each organism is determined by the ability of such organism to adapt itself to its environment.

He established these revolutionary theories in his book called **"On the Origin of Species" (1859)**.

The theory of evolution of Charles Darwin is based on five key observations and the deductions learned from these.

These observations and deductions are summarized by the great biologist **Ernst Mayr** as follows:

1) Species have great fertility.

2) Populations remain approximately about the same size, with modest fluctuations.

3) The food supplies are limited, but are relatively constant the most part of the time.

From these three observations, it can be said that there will be a struggle for the survival among individuals.

4) In sexual reproduction, generally two individuals are not identical. The variation is extensive.

5) And, many of these variations are inherited. From this we can infer that:

In a world of stable populations where each individual must struggle to survive, those with the **best characteristics** will be the most likely to survive; and those advantageous traits (or characteristics) will be passed to their offspring.

These advantageous characteristics are inherited by the following generations, becoming predominant among the population through time. **This is Natural Selection.**

It can also be inferred that natural selection, if carried far enough, will make changes in the population, eventually leading new species.

"These observations have been widely demonstrated in biology, and even fossils show the veracity of these observations".

The Evolution Theory of Darwin

- **Variation:** There is a variation in each population.

- **Competition:** Organisms compete for limited resources.

- **Procreation:** Organisms procreate more than what they can live.

- **Genetics:** Organisms transcend genetic traits to their offspring.

- **Natural Selection:** Those organisms with the most beneficial traits are more likely to survive and reproduce.

After the publication of "On the Origin of Species", the biography of Charles Darwin tells us that he continued writing about botany, geology, and zoology until his death in 1882.

He is buried in Westminster Abbey, England.

APPENDIX #4

THE MANIFESTATION OF LOVE

There is a force that moves humanity. It is more powerful than electricity, conquers more than money; and attracts more than gravity:

IT IS THE POWER OF LOVE

It comes from the primary source of all things: God.

His Word tells us that God wants to show His love to each person. This includes YOU. He says to you:

> "I drew them with bands of love." (Hosea 11:4)

These bands speak of a sacrifice, for love, in order to save YOU.

Perhaps your ways are far from God; however, He seeks to save, bless, prosper; and above all, to have communion with you.

> "For God so loved the world that he gave his only begotten Son, that whosoever believeth in him should not perish, but have everlasting life." (John 3:16)

> "But God commendeth his love toward us, in that, while we were yet sinners, Christ died for us." (Romans 5:8)

Draw closer to God. He is your creator. Let Him save you. Do not resist the call of His love.

It is very easy to receive the forgiveness of God, because Christ already paid the price that God required.

CARRY OUT NOW WITH YOUR PART:

a) Recognize that you are a sinner and that you have not lived perfectly before God.

b) Repent sincerely of living away from God.

c) Ask Him for forgiveness from your heart for all your sins.

d) Receive the grace of God in your life, confessing Jesus Christ as your Lord and Saviour.

e) Learn to live according to the will of God. For this, it is necessary that you learn to talk with God (pray); learn what God wants you to know (read the Bible, congregate in a church where the Word of God is taught and preached).

f) Share with others the love you have received.

The Lord says:

"Come now, and let us reason together, saith the LORD: though your sins be   as scarlet, they shall be as white as snow; though they be red like crimson,   they shall be as wool." (Isaiah 1:18)

"That if thou shalt confess with thy mouth the Lord Jesus, and shalt believe in thine heart that God hath raised him from the dead, thou shalt be saved." (Romans 10:9)

"If we confess our sins, he is faithful and just to forgive us our sins, and to cleanse us from all unrighteousness." (1 John 1:9)

"And the blood of Jesus Christ his Son cleanseth us from all sin." (1 John 1:7b)

Pray to God, ask Him to forgive all your sins and receive His grace and love.

APPENDIX #5

Your children and evolution.

Suggestions for Christian parents [66]

By: Geoff Chapman

February 25, 2008

Let's be honest. In the story of the king's new clothes, it was a boy who saw that the king had no clothes on at all. Even if parents do not have the courage to acknowledge it, **children quickly recognize that evolution contradicts the Bible and undermines the Christian faith.**

How can parents encourage their children to share their faith and maintain their trust in the Bible? I suggest the following activities:

- Teach your children about God's Creation starting from when they are very young. Read the Biblical narration over and over again until they have it memorized.

Read to them about God's wonderful creatures from illustrated books, but make sure they don't contain any theistic evolution, progressive creation, and day-age or gap theory. Avoid so-called "Christian" books which begin with the world as a *melted blob*, or that include *"millions of years ago"*.

- Instill into your children a sense of wonder towards the greatness and magnificence of what God has done (while remembering that the world we see today has been affected by the Curse).

Whenever you go out into the country, or even into a garden, show them the wonders that can be observed.

Teach them what Genesis talks about: the birds, the bats, the flying insects, and constantly remind them that *God made the birds fly on the fifth day.*

Take them outside at night and show them the moon that *God created on the fourth day.*

Show them the detailed patterns in leaves and *remind them over and over again how absurd* it is to suggest that such pretty patterns happened by accident.

- Teach your children about the fallacies of evolution **before** they learn it in school. Make sure that you as a parent know the mistakes about the evolutionary theory.

To do this, you need to work and study good books on the subject. Then you will have to work by teaching these facts to your children in your home.

- Since dinosaurs are used so often, even from a kindergarten level, to introduce children to evolution, remind them that terrestrial dinosaurs were made on the *sixth day* of Creation (and other dinosaur like creatures, such as; Plesiosaurs, on the fifth day).

Inform them about the evidence about the dinosaurs that lived with people, about the rock carvings, and the dragon legends. Always be ready to point out when a commentator is talking about evolution.

- Encourage them to be saddened by the fact that so many people believe what is not true and substitute chance for the wonderful gifts of God.

- Encourage your children to study the real evidence for themselves. Take them to places where they can find fossils and explain how fossils are usually formed by rapid burial in sediment.

Explain to them how the great Flood in Genesis provided the perfect conditions for the formation of millions of fossils around the Earth. Point out that the fossils in sedimentary rocks do not show that the rocks formed slowly and that the Earth is old. To the contrary, they indicate rapid processes and therefore a *recent catastrophe by water* as the Bible teaches.

Mention to them that animals that die today, are very unlikely to become fossils, since this does not happen daily.

Finally, explain to them in detail why there is suffering and cruelty in the world. Tell them where death comes from and its significance.

Present to them how God formed a perfect world but the sin of Adam spoiled that perfection and brought death and decay to the Earth.

Tell them that Adam was punished with the curse of death, and the God of Creation knew that He himself would come in the form of Jesus Christ, the last Adam, to suffer the same curse of death himself.

Then, all those who accept the sacrifice for sin will have forgiveness and eternal life, Jesus restored the relationship that was broken with God, and that at the end of time, God will create a new heaven and a new earth-a completely restored creation in which all those who truly love Him will be able to share in peace and harmony.

Bibliography

Albalat, Indalecio Gil. ¿A Dónde va la Tierra?. Editorial CLIe. España. 1990

Animal Pharming: The Industrialization of Transgenic Animals. December 1999. Center for Emerging Issues.

Astakhoff, Saloff. Origen y Destino del Planeta Tierra. Editorial CLIE. España. 1983.

"El Ciclo Hidrológico (Panfleto), U.S. Geological Survey, 1984"

Ferrell, Vance. The Evolution HandBook. Evolution. Facts Inc. USA. 2005

Figueroa, Rocío A. La verdad y el conocimiento científico. Editorial FIGARO. 2007

Freeman, Scott; Herron, Jon. Análisis Evolutivo. Segunda Edición. Prentice Hall, Madrid, España. 2002

Gish, Duane T. Creación, Evolución y el Registro Fósil. Libros CLIE. España. 1988

Huse, Scott M. El Colapso de la Evolución. Chick Publications. USA. 1993

Kupelian, David. The marketing of the evil. Cumberland House Publishing, Inc. 2005

Lester, Lane P. & Hefley, James C. Clonación Humana. Editorial Portavoz. USA. 2000

Mathews, Van Holde & Ahern. Bioquímica. Tercera Edición. Pearson Educación, S.A., Madrid, España. 2002.

McKee Trudy and McKee James. Bioquímica. La base molecular de la vida. Tercera Edición. McGraw Hill Interamericana de España. SAV

McDowell, Josh. Evidencia que exige un veredicto. 10ma. Impresión. Editorial Vida. 1995.

Morris, H.M. Geología, ¿Actualismo o Diluvialismo?. Libros CLIE. España. 1983

Muñoz, Nahum. Génesis, Desde la Creación hasta Abraham. Trinity Church Int'l. USA. 2000

Ouweneel, Willem. Biología y Orígenes. Editorial CLIE. España. 1989

Pardo, Antonio. Departamento de Humanidades Biomédicas; Facultad de Medicina, Universidad de Navarra, Pamplona. SCRIPTA THEOLOGICA 39 (2007/2) 551-572. ISSN 0036-9764

Ross, Hugh. El Creador y el Cosmos. Editorial Mundo Hispano. USA. 1999

Se descubrió que No somos tan parecidos al chimpancé. Gaceta Universitaria. Ciencia y Tecnología. González de Alba, Luis.

Solomon, Eldra; Berg, Linda; y Martin, Diana. Biología. McGraw Hill Int'l. 2001

Starr, Cecie & Taggart, Ralph. Biología. La unidad y diversidad de la vidad. Décima edición. Internacional Thomson Editores, S.A. 2004

Vila, Samuel. A Dios por el Átomo. Editorial CLIE. España. 1987

Antonio Pardo. Departamento de Humanidades Biomédicas; Facultad de Medicina, Universidad de Navarra, Pamplona. SCRIPTA THEOLOGICA 39 (2007/2) 551-572. ISSN 0036-9764

Web links

ADN Y GENETICA.
http://aula2.elmundo.es/aula/laminas/lamina1170929412.pdf

Cazau, Pablo. La Teoría del Caos. 2002
http://www.antroposmoderno.com/antro-articulo.php?id_articulo=152

Chapman, Geoff. Sus hijos y la evolución. Sugerencias para padres Cristianos. February 25, 2008.
www.answersingenesis.org/sp/articles/cm/v6/n4/evoluti on

Charles Darwin y El origen de las especies
http://redescolar.ilce.edu.mx/redescolar/act_permanentes/historia/histdeltiempo/mundo/prehis/t_teoesp.htm

Ciencia, Evolución o Creación.
http://www.slideshare.net/Jonshuan/ciencia-evolucion-o-creacion

Creación o evolución ¿Importa realmente lo que creamos? publicación de la Iglesia de Dios Unida, ESTADOS UNIDOSP.O. Box 541027Cincinnati, OH 45254-1027Sitio en Internet: www.ucg.org

CREACIÓN VERSUS EVOLUCIÓN.
http://home.coqui.net/apoc7/AR-Creacion-Evolucion.htm

Creacionismo vs. Evolucionismo.
http://members.tripod.com/~Seresma/Spanish_CvsE.html
Crean ratón con un cromosoma humano.
http://axxon.com.ar/not/154/c-1540237.htm

Creationism 'no place in schools'
http://news.bbc.co.uk/1/hi/education/4896652.stm

CROMOSOMAS http://www.iqb.es/cancer/g006.htm
Datos interesantes del ADN

Descifran cromosoma 22 del chimpancé
El Tiempo. Mayo 27 de 2004.
http://www.abacolombia.org.co/postnuke/modules.php?op=modload&name=News&file=article&sid=269

Descripción general del sistema vascular.
http://www.healthsystem.virginia.edu/uvahealth/adult_cardiac_sp/overvasc.cfm

Dióxido de Carbono y Metabolismo.
http://homepage.mac.com/uriarte/metabolismo.html&h=
353&w=633&sz=44&hl=en&start=2&um=1&usg=__G
JzroUCi2_kgMqUwRG1oX93w0Uo=&tbnid=E3M3T

El ADN y los seres humanos.
http://eleccionesdominicanas.com/2008/04/02/politica-
conciencia-y-autenticidad/1561/Publicado el 2-04-2008
por José R. Bourget Tactuk

El contenido del semen
http://www.sexologia.com/index.asp?pagina=http://ww
w.sexologia.com/articulos/semen/elsemen.htm

El cromosoma 21: Anotación funcional.
http://www.down21.org/salud/genetica/cromosoma21.htm

El genoma de la gallina. http://waste.ideal.es/genoma-
gallina.htm

El genoma del ratón, la clave de la investigación
biomédica. http://waste.ideal.es/genoma-raton.htm

El Gran Colisionador De Hadrones LHC (jananet)
http://www.slideshare.net/jananet/hadrones-
presentation/

El juicio del mono. www.tecnociencia.org/n/380/juicio-
mono/&h=149&w=190&sz=9&hl=en&start=12&um=1
&tbnid=dr_BY7-
UNUM8CM:&tbnh=81&tbnw=103&prev=/images%3F
q%3Djuicio%2Bdel%2Bmono%26um%3D1%26hl%3D
en%26rlz%3D1T4GZHZ_enUS223US224%26sa%3DN

El Misterio de los Dinosaurios por fin revelado.
http://www.antesdelfin.com/misteriodinosaurios.html

EL ORIGEN DE LA VISIÓN.
http://www.uam.es/personal_pdi/psicologia/travieso/we
b_percepcion/sistemav.html&h=240&w=384&sz=74&h
l=en&start=7&um=1&usg=__HfL7PyckeJmu6QpX2W
TgTFnmox8=&tbnid=zqUEkpImUd4IzM:&tbnh=77&tb
nw=123&prev=/images%3Fq%3Dojo%2BESCLEROTI
CA%26um%3D1%26hl%3Den%26rlz%3D1T4GZHZ_
enUS223US224%26sa%3DN

El Sistema Nervioso central
http://www.monografias.com/trabajos11/sisne/sisne.shtml

El sistema reproductivo de la mujer.
http://www.educasexo.com/files/media/2-aparato-
reproductor-
masculino.jpg&imgrefurl=http://www.educasexo.com/a
dolescentes/el-sistema-reproductivo-del-
hombre.html&h=345&w=275&sz=42&hl=en&start=15
&um=1&usg=__HF5VjH2IUMSiniZxnWvvyb3EDgM=
&tbnid=kcy_M7WxyrxQ8M:&tbnh=120&tbnw=96&pr
ev=/images%3Fq%3Dsistema%2Breproductivo%26um
%3D1%26hl%3Den%26safe%3Doff%26rlz%3D1T4GZ
HZ_enUS223US224%26sa%3DN

El sistema reproductivo del hombre.
http://www.educasexo.com/adolescentes/el-sistema-
reproductivo-del-
hombre.html&h=345&w=275&sz=42&hl=en&start=15
&um=1&usg=__HF5VjH2IUMSiniZxnWvvyb3EDgM=
&tbnid=kcy_M7WxyrxQ8M:&tbnh=120&tbnw=96&pr
ev=/images%3Fq%3Dsistema%2

Ensayo sobre la Teoría del Caos y la visión de Dee Hock.
http://www.monografias.com/trabajos13/caos/caos.shtm
l

ESTRUCTURA Y COMPOSICION DE LA MATERIA. http://hnncbiol.blogspot.com/2008/01/el-estado-en-que-se-encuentra-la_3800.html

Evolution isn't enough, professor says.
http://www.msnbc.msn.com/id/9729036/

Evolution Less Accepted in U.S. Than Other Western Countries, Study Finds
http://news.nationalgeographic.com/news/bigphotos/213
29204.html
Evolution: Facts, Fallacies and Implications
http://www.thercg.org/books/effai.html?cid=g0207&s_k
wcid=proofs%20of%20evolution|951159121&gclid=CJ
awjP2ej5ECFQ2nGgod622EGQ
EXPLORING EVOLUTION
http://science.nsta.org/enewsletter/2003-
11/member_high.htm

FISIÓN NUCLEAR
http://erenovable.com/2006/06/01/fision-nuclear/
FUNDAMENTOS DE TERMODINAMICA
Y TERMOQUIMICA Marzo 15, 2008 — Juan José

Gallinas y humanos comparten un 60% de sus genes
http://www.elpais.com/articulo/sociedad/Gallinas/huma
nos/comparten/genes/elpepusoc/20041208elpepusoc_5/
Tes

Genoma Humano
http://www.monografias.com/especiales/genoma/index.s
html

Hombre y chimpancé 96% de ADN parecido
http://www.terra.com.mx/articulo.aspx?articuloid=1692
20&paginaid=1
http://labquimica.wordpress.com/2008/03/15/fundament
os-de-termodinamica-y-
termoquimica/&h=367&w=566&sz=29&hl=en&start=9
2&um=1&usg=__vI8kPRrNVuWKhy3DB_VmgbXnLd
g=&tbnid=JDRlJ6Gx93wW3M:&tbnh=87&tbnw=134&
prev=/images%3Fq%3Dlas%2Bleyes%2Bde%2Bla%2B
Termodin%25C3%25A1mica%26start%3D80%26ndsp
%3D20%26um%3D1%26hl%3Den%26rls%3Dcom.mic
rosoft:en-us:IE-
SearchBox%26rlz%3D1I7GZHZ%26sa%3DN

Kukso, Federico . Biografía no autorizada del gusano
más famoso (y menos reconocido) de la ciencia
http://www.nacionapache.com.ar/archives/1540

La Biblia By Matthew J. Slick
http://www.lasperseveradoras.org/es/verdadesbiblicas/L
aBiblia/AcercadelaBiblia.cfm

La Clonación: de los animales al hombre
http://www.todo-
ciencia.com/reportaje/0i34158700d990138942.php

La secuencia del genoma del chimpancé muestra que
comparte un 96% con el humano

http://www.elmundo.es/elmundosalud/2005/08/31/bioci
encia/1125506184.html

Leyes de la Probabilidad
http://www.uaq.mx/matematicas/estadisticas/xu4.html#t4

Lo esencial está en los genes
http://waste.ideal.es/genesevolucion.htm

Mapa Genético Humano: Genoma
http://www.bolivia.com/especiales2003/genoma/notas/gl
osario.asp

MENSAJE DEL SANTO PADRE JUAN PABLO II
A LOS MIEMBROS DE LA ACADEMIA
PONTIFICIA DE CIENCIAS
http://www.vatican.va/holy_father/john_paul_ii/message
s/pont_messages/1996/documents/hf_jp-
ii_mes_19961022_evoluzione_sp.html

National Science Teachers Association.
Nociones básicas de Citología: La
mitosis celularhttp://permian.wordpress.com/2008/08/13
/nociones-basicas-de-citologia-la-mitosis-
celular/&h=489&w=614&sz=64&hl=en&start=8&um=
1&usg=__ATEw5EGlHz4_6k90Uzt58Q8hzGo=&tbnid
=E5bGO-
7GeUtDIM:&tbnh=108&tbnw=136&prev=/images%3F
q%3Dmaterial%2Bgen%25C3%25A9tico%26um%3D1
%26hl%3Den%26safe

PEIRCE Y LA TEORÍA DEL CAOS. McNabb, Darin
Costa Instituto de Filosofía, Universidad Veracruzana,
México dcosta@uv.mx

Poll: Majority Reject Evolution
http://www.cbsnews.com/stories/2005/10/22/opinion/pol
ls/main965223_page2.shtml

Pope criticizes atheism, modern Christianity, in
encyclical on hope
http://www.iht.com/articles/ap/2007/11/30/europe/EU-
GEN-Vatican-Encyclical.php?WT.mc_id=rsseurope

http://ga.water.usgs.gov/edu/watercyclespanish.html

Powell, Jacqueline Melissa. EVOLUTION EXPOSED!
OPPOSING SCIENCE AND SCRIPTURE. May 1994.
A Research Paper Presented to The Faculty of the
English Department Tabernacle Baptist Bible College

PROBLEMAS CIENTIFICOS CON LA TEORIA DE
LA EVOLUCION DE LAS ESPECIES
http://www.antesdelfin.com/problemasdelaevolucion.ht
ml

Problemas Con La Macro-evolución
http://www.iglesiabautista.org/articulos/view/?id=24

Prohíben teorías de Darwin en Serbia
http://www.elsiglodetorreon.com.mx/noticia/109552.pro
hiben-teorias-de-darwin-en-serbia.html

Questions and Answers on the teaching of Evolution.
www.nsta.org/pdfs/EvolutionQandA.pdf

Second Law of Thermodynamics - Does this basic law
of nature prevent Evolution?

http://www.christiananswers.net/q-eden/edn-thermodynamics.html

Semen http://es.wikipedia.org/wiki/Semen

Shisher, Harold S. y Whitelaw, Robert L. Las Dataciones Radiométricas-CRITICA-. Libros CLIE. España. 1980

Sistema endocrino
http://www.profesorenlinea.cl/imagenciencias/sistemaen docrino002.jpg&imgrefurl=http://www.profesorenlinea. cl/Ciencias/sistemaEndocrino.htm&h=401&w=534&sz= 86&hl=en&start=5&um=1&usg=__HyfojybbLyG-

Termodinámica
http://www.jfinternational.com/mf/termodinamica.html

Termodinámica
http://www.monografias.com/trabajos/termodinamica/te rmodinamica.shtml

Vélez, Antonio. Ed. Univ.de Antioquia, Medellín,1994. http://lloro.galeon.com/E%20V%20O%20L%20U%20C %20I%20O%20N.htm

http://www.escolar.com/cnat/08GRANDE.gif&imgrefur l=http://www.escolar.com/cnat/08sisnerv.htm&usg=__O kn05e-
AEmg6hXY4plfplBobRNY=&h=354&w=293&sz=30& hl=en&start=5&um=1&tbnid=pOO3FVQg0kuUnM:&tb nh=121&tbnw=100&prev=/images%3Fq%3Dsistema% 2Bnervioso%26um%3D1%26hl%3Den%26rlz%3D1T4 GZHZ_enUS223US224%26sa%3DN

http://www.secundariasgenerales.tamaulipas.gob.mx/An
atom%25EDa/sistema%2520muscular.jpg&imgrefurl=h
ttp://www.secundariasgenerales.tamaulipas.gob.mx/Ana
tom%25EDa/muscular.htm&usg=__ZdGBRgUjouz-
FOkjuh_BycUQ6es=&h=441&w=640&sz=130&hl=en
&start=22&um=1&tbnid=dgLjh6RNsgiCwM:&tbnh=94
&tbnw=137&prev=/images%3Fq%3Dsistema%2Bmusc
ular%26start%3D20%26ndsp%3D20%26um%3D1%26
hl%3Den%26rlz%3D1T4GZHZ_enUS223US224%26sa
%3DN

http://medlineplus.gov/

http://upload.wikimedia.org/wikipedia/commons/thumb/
d/df/Esqueleto_humano_(vista_frontal).svg/311px-
Esqueleto_humano_(vista_frontal).svg.png&imgrefurl=
http://commons.wikimedia.org/wiki/Image:Esqueleto_h
umano_(vista_frontal).svg&h=599&w=311&sz=102&tb
nid=xjJzN_Woub0J::&tbnh=135&tbnw=70&prev=/ima
ges%3Fq%3Desqueleto%2Bhumano&usg=__1ZIm30O
SczEElubW57CNVVGF1mw=&sa=X&oi=image_result
&resnum=1&ct=image&cd=1

http://mx.encarta.msn.com/media_461547449_7615606
28_-
1_1/Sistema_nervioso_aut%25C3%25B3nomo_o_veget
ativo.html&usg=__yu06lANl4kunt28Bd3TRgyb0rOs=
&h=328&w=556&sz=28&hl=en&start=18&um=1&tbni
d=ShbL2IlUpOCteM:&tbnh=78&tbnw=133&prev=/ima
ges%3Fq%3Dsistema%2Bnervioso%2Bperiferico%26u
m%3D1%26hl%3Den%26rlz%3D1T4GZHZ_enUS223
US224

http://www.secundariasgenerales.tamaulipas.gob.mx/An
atom%25EDa/sistema%2520hormonal.jpg&imgrefurl=h
ttp://www.secundariasgenerales.tamaulipas.gob.mx/Ana
tom%25EDa/endocrino.htm&usg=__JCx2sooN8KEnLd
4bnBEm3e03xiQ=&h=441&w=640&sz=162&hl=en&st
art=1&um=1&tbnid=U9K5rvR9-Prn-
M:&tbnh=94&tbnw=137&prev=/images%3Fq%3Dsiste
ma%2Bhormonal%26um%3D1%26hl%3Den%26rlz%3
D1T4GZHZ_enUS223US224

http://2.bp.blogspot.com/_8NtbI0QpLkw/SLAqO0lUGlI
/AAAAAAAAACU/IkyFrt2_LTM/s200/fecundacion-
NTnva.jpg&imgrefurl=http://bioeticaylibros.blogspot.co
m/&usg=__p9_Mcz8Uh7RmqpY5uYtVKiOp4q4=&h=
146&w=200&sz=9&hl=en&start=87&um=1&tbnid=jXi
t-
7NMI_KxPM:&tbnh=76&tbnw=104&prev=/images%3
Fq%3Dfecundacion%2Besperma%2Bcantidad%26start
%3D80%26ndsp%3D20%26um%3D1%26hl%3Den%2
6rlz%3D1T4GZHZ_enUS223US224%26sa%3DN

http://www.thefreedictionary.com/interpretational

References

[1] All of the biblical verses have been taken from the King James Bible version. When another version is used, the source will be indicated.

[2] Atheists celebrate Darwin Day in Coconut Creek by LOIS K. SOLOMON-South Florida Sun-Sentinel http://www.sunsentinel.com/news/local/broward/sfl-flbdarwinday0204brfeb04,0,5555923.story

[3] Educators say evolution still a "theory". Panel had asked state board to teach ideas as fact http://www.wnd.com/index.php?fa=PAGE.view&pageId=56830

[4] Imagen de: http://www.yoism.org/images/BigBang2.JPG

[5] http://www.slideshare.net/jananet/hadrones-presentation/

[6] http://www.astrored.net/origen_del_universo/

[7] SEE: *evolutionibus.eresmas.net/ciencia.html*

[8] Solomon, Eldra; Berg, Linda; Martin, Diana. BIOLOGIA. McGraw Hill Internacional. 2001. Página 374

[9] http://en.wikipedia.org/wiki/Telescope
[10] http://en.wikipedia.org/wiki/Hubble_Space_Telescope
[11] Folleto: ¿Creación o evolución? ¿Importa realmente lo que tú crees? S-CE/05-2005/1.0 (pág. 5) Iglesia de Dios Unida. U*na Asociación Internacional.* www.ucg.org

[12] Op. Cit. Pág. 1

[13] Op. Cit. Pág. 2; William Federer, America's God and Country ["El Dios y el país de los Estados Unidos"], 1996, p. 61.

[14] Op. Cit. Pág. 5

[15] http://www.scoop.co.nz/stories/HL0803/S00051.htm

[16] Antonio Pardo. Departamento de Humanidades Biomédicas; Facultad de Medicina, Universidad de Navarra, Pamplona. SCRIPTA THEOLOGICA 39 (2007/2) 551-572 .ISSN 0036-9764

[17] Dennis O'Neil, Early *Theories of evolution*

[18] Tarr-Taggart. Biología. 10ma. Edic. Pagina 11

[19] I.L. Cohen, *Darwin Was Wrong A Study in Probabilities* (P.O. Box 231, Greenvale, New York 11548: New Research Publications, Inc., 1984), p. 205.

[20] A.J. White, "Uniformitarianism, Probability and Evolution," *Creation Research Society Quarterly*, Vol. 9, No. 1 (June 1972), pp. 32-37.

[21] Mathematician Emil Borel agrees that the laws of probability demonstrate that: *"Events whose probabilities are extremely small never occur."*

[22] Fred Hoyle and N. Chandra Wickramasinghe, Evolution from Space (Aldine House, 33 Welbeck Street, London W1M 8LX: J.M. Dent & Sons, 1981), p. 148, 24, 150, 30, 31

[23] McKee Trudy and McKee James. Bioquímica. La base molecular de la vida. Tercera Edición. McGraw Hill Interamericana de España. SAV

[24] Puedes leer todos los pormenores del estudio: "PROBLEMAS CIENTIFICOS CON LA TEORIA DE LA EVOLUCION DE LAS ESPECIES" en www.antesdelfin.com/problemasdelaevolucion.html

[25] Robert Shapiro, (Ph.D.), Origins: A Skeptics Guide to Creation of Life on Earth (Simon & Schuster, 1986), pp.98-117.

[26] Charles Thaxton (Ph.D. Chemistry), Walter Bradley (PhD. Material Science), Roger Olsen (Ph.D. Geochemistry), The Mystery of Life's Origins: Reassessing Current Theories (New York: Philosophical Library, 1984), p.66.

[27] p. 292, first paragraph of Chapter 9, "On the Imperfection of the Geologic Record", of The Origin of Species.

[28] David Raup (Ph.D. Harvard University), "Conflicts Between Darwin and Paleontology", Field Museum of Natural History, Vol. 50, No. 1 (January 1979) p.22.

[29] Fred Hoyle and C. Wickramasinghe, Evolution >From Space (London: J.M. Dent & Sons, 1981), p. 8,70.

[30] Fred Hoyle and C. Wickramasinghe, Evolution >From Space. pp. 148,24,150,30,31

[31] Hubert Yockey, Ph.D., Information Theory and Molecular Biology, (Cambridge University Press, 1992), p.257.

[32] Francis Crick and L.E. Orgel (1973), "Directed Panspermia", Icarus, 19: 341-346.

[33] Michael Denton, Evolution, pp.326-328.

[34] H.J. Muller, "How Radiation Changes the Genetic Constitution", Bulletin of the Atomic Scientists Vol.11, No.9 (November 1955), p.331 (emphasis added).

[35] Pierre Paul Grasse, PhD., Evolution of Living Organisms (New York: Academic Press, 1977) pp.88,103.

[36] McKee Trudy and McKee James. Bioquímica. La base molecular de la vida. Tercera Edición. McGraw Hill Interamericana de España. SAV. Pág. 1

[37] www.elnuevoherald.com/noticias/sur-de-la-florida/story/162781.html

[38] Adaptado de: http://www.secundariasgenerales.tamaulipas.gob.mx/

[39] Adaptado de: http://www.escolar.com/

[40] Adaptado de www.healthsystem.virginia.edu/uvahealth/adult_cardiac_sp/overvasc.cfm

[41] http://www.iglesiabautista.org/articulos/view/?id=24

[42] http://news.bbc.co.uk/hi/spanish/science/newsid_6124000/6124174.stm

[43] Animal Pharming: The Industrialization of Transgenic Animals December 1999
[44] http://waste.ideal.es/genoma-gallina.htm

[45] www.clarin.com/diario/2007/01/14/um/m-01345070.htm

[46] www.absurddiari.com/s/llegir.php?llegir=llegir&ref=7903

[47] CLONACIÓN, RECORRIDO CRONOLÓGICO por Javier de Rios Briz. http://www.todo-ciencia.com/reportaje/0i34158700d990138942.php

[48] (Creation and Evolution: Rethinking the Evidence From Science and the Bible, 1985, p. 80). ("La creación y la evolución: Nuevo análisis de los hechos de la ciencia y de la Biblia")

[49] Starr-Taggart. "Biología, la unidad y diversidad de la vida" 10ma. Edición. Editorial Thomson, Página 15

[50] Creación o Evolución. ¿Importa realmente lo que creamos? Iglesia de Dios Unida. ESTADOS UNIDOS P.O. Box 541027Cincinnati, OH 45254-1027Sitio en Internet: www.ucg.org

[51] Op. Cit. Pag. 27

[52] http://wnd.com/index.php?fa=PAGE.view&pageId=95913

53

http://wnd.com/index.php?fa=PAGE.view&pageId=902
41

54

www.elsiglodetorreon.com.mx/noticia/109552.prohiben
-teorias-de-darwin-en-serbia.html

[55] Tomado de: La Biblia, By Matthew J. Slick 1998, 2000. MINISTERIO DE APOLOGETICA

[56] LA SINGULARIDAD DE LA BIBLIA. Recopilado por Damián, USA.

[57] Análisis más recientes dan 5.730 años como mejor «vida media», lo que representa un error de sólo un 3 %.

[58] 16,2 en conchas marinas, y 15,3 en vegetación y tejidos vivientes, debido a la diferente relación de C12/C14 en cada grupo.

[59] W. F. Libby (1952). *Radiocarbon dating.* University of Chicago Press, 1st edition. Publicado en castellano, *Datación Radiocarbónica* (Ed. Labor, Barcelona, 1970).

[60] *Ibid.* (1955), 2nd ed, (con una lista ampliada de dataciones mediante C14 en el capítulo 6, y la adición del capítulo 7 por F. Johnson, titulado: «Reflections upon the significance of radiocarbon dates»).

[61] See *Crítica de las Dataciones Radiométricas*, de H. S. Slusher, Colección Creación y Ciencia n° 3 (SEDIN/Clie, Terrassa, España 1980).

[62] L. S. B. Leakey (1959). A new fossil skull from Olduvai, *Nature*, 184:491.

[63] F. C. Howell, 1969. Remains of hominidae from pliocene and pleistocene formations in the lower Omo basin, Ethiopia, *Nature*, 223:1234.

[64] *Datelines in Science*. November 7, 1969, 1,5 million years are added to early hominids' age. Véase también *Datelines in Science*. september 17, 1967, sobre el cráneo de la Garganta de Olduvai de la Referencia 5.

[65] http://www.galapagos-islands-tourguide.com/biografia-de-charles-darwin.html

[66]

http://www.answersingenesis.org/sp/articles/cm/v6/n4/evolution